All Aces, No Jokers

All Aces, No Jokers

The Wartime Memoirs of an American Fighter Pilot

Thomas Littleton, PhD

Lieutenant Colonel, USAF (Ret.)

Amazon Hardback Edition ISBN: 979-8-9851875-0-2
Amazon Paperback Edition ISBN: 979-8-9851875-1-9
Barnes & Noble Paperback Edition ISBN: 979-8-9851875-4-0
eBook Kindle Edition ISBN: 979-8-9851875-2-6
eBook Nook Edition ISBN: 979-8-9851875-3-3

2nd Edition

The views expressed in this publication are those of the author and do not necessarily reflect the official policy or position of the Department of Defense or the U.S. government.

Preface

It was February 17th, 2018, and I was sitting at a table in a large ballroom at the Treasure Island in Las Vegas. The room was filled with people, drinking, and laughing as they ate dinner. The occasion was the 78th Bushmaster's Centenary, the 100th Anniversary of the founding of one of the oldest squadrons in the U.S. Air Force. Originally assembled in Texas as the 78th Aero Squadron in 1918, the unit was stationed in the Panama Canal Zone during the 1930s. It was there they picked up the name, "Bushmasters." The "Bushmaster" is a kind of rattlesnake indigenous to the jungles of Panama. Bushmasters are silent rattlesnakes, large, territorial, lethal, aggressive, and known to strike without warning; just like the pilots of the 78th! The 78th Fighter Squadron played an interesting part in World War II. On December 7th, 1941, a

78th pilot was the guy who, upon learning of a large formation of airplanes on radar approaching the Hawaiian Islands famously dismissed the Japanese planes as friendly airplanes flying in from the States, missing the opportunity to raise the alarm of the impending attack. The squadron was soon wiped out on the ground at Wheeler Field by the Japanese sneak-attack. The Bushmasters rebuilt and eventually flew the very last combat mission of World War II. The pilot who led that mission, Captain Jerry Yellin, was scheduled to be our guest speaker for the Bushmaster Centenary, but sadly he passed away just a few weeks before the big celebration. After World War II the 78th Fighter Squadron continued to play a major role in securing America's freedom during the Cold War. In the 1990's the Bushmasters moved from England to Shaw Air Force Base in South Carolina, where they played a pivotal role in the enforcement of the No-Fly Zones over Iraq. In 1999 the squadron deployed to Aviano Air Base in Italy to support of Operation Allied Force, the NATO air campaign against the Federal Republic of Yugoslavia to stop the ethnic cleansing in Kosovo. The 78th went on to distinguish itself in the post-9/11 years fighting terrorism, and today proudly continues the legacy of those who flew before them.

I joined the 78th Fighter Squadron in early 1996, flying F-16 Fighting Falcons, more commonly called the "Viper" by the men and women who fly and maintain them. The F-16CJs I flew at Shaw Air Force Base were the most advanced variant of the F-16 in service with the U.S. Air Force. The F-16CJ is a single-seat, single engine, multi-role, all-weather fighter. Capable of flying at more than twice the speed of sound, possessing 9-G maneuverability and armed with the most lethal weapons ever devised, the F-16CJ was the very tip of Democracy's spear. The F-16CJ was configured with special equipment and weapons to detect, locate, target, and destroy an enemy's surface to air missiles (SAMs) and anti-aircraft artillery (AAA). This particular kind of capability is called Suppression of Enemy Air Defenses, or SEAD (pronounced, "SEED"). Pilots flying the SEAD mission are called "Wild Weasels", and SEAD is hands-down the most challenging and dangerous mission ever flown by fighter pilots. As a Wild Weasel your job is to get the enemy to try and shoot you down, and in doing so expose their locations so you can kill them first, thus clearing the target area for the people who need to put good-guy warheads on bad-guy foreheads. At the same time, the jet's powerful radar and compliment of air-to-air missiles allow it to defeat enemy fighter planes as well. When this air-to-air capability is combined with the jet's SEAD capability, F-16CJs provide what the Air Force calls "Force Protection." Wild Weasels pride ourselves on being "first-in and last-out"; we are the first ones flying into the target area and the last ones to leave. A Weasel's job is to protect the strike package against all threats: SAMs, AAA (also called "flak"), and enemy fighter planes. Nobody flies into enemy airspace without an escort of Wild Weasels for protection. The mission is so dangerous and well-respected by other military pilots that Weasels always drink for free at the Officer's Club on a Friday night!

Sitting at the table with me that night were several pilots with whom I had flown F-16CJs in the 78th Fighter Squadron. The conversation flowed freely, as our wives talked, and the pilots joked with one another and told "remember when" stories of flying fast jets. A lot of the stories were from our time flying out of Italy during the Kosovo Air War. In the middle of all this I glanced around the table, and I caught the eye of two of my old squadron mates, Chewy and Goose. For a few seconds the three of us looked back and forth at each other rather seriously, and in that moment, I knew exactly what we were all three thinking; of all the pilots in this room, we three had flown combat together. We had flown and fought beside one another during Operation Allied Force, the Kosovo Air War...or as we used to joke, "World War Yugo-Nam.". Pilots everywhere share the comradery of service and a love of flying, but there's a special bond between fighter pilots who have flown combat together. There's a connection there I have never found anywhere else. The respect comes from having literally trusted each other with our lives. In aerial combat, teams win, and individuals lose. You depend on each other to kill and survive when chaos reigns in the fog of battle. There's no way to describe what it feels like to climb out of the jet after a hard-fought mission and know that good guys lived, and bad guys, died because of that teamwork. Looking at each other, Chewy, Goose and I were all thinking the same thing, "I checked that guy's six, and he checked mine." Chewy nodded his head a few times, silently acknowledging what we were all thinking. Goose raised his beer a few inches off the table, and Chewy and I tipped ours back to him in response. The moment passed, but what it represented would never be forgotten. So, I decided to write a book about it...what it was like to be a fighter pilot, flying combat missions in the great Kosovo Air War, and living the adventure of a lifetime.

I do want to take a second and explain some things up front. The events I describe are relayed to the best of my memory; I take full responsibility for any mistakes or inaccuracies. Getting old sucks, and every day I remember less. I also made sure than nothing I describe in the book will tip-off our nation's adversaries to the tactics, techniques, and procedures used by today's pilots. Additionally, I've added conversations between people to make the events more realistic and interesting to read. Also, with very few exceptions, I have changed the names of the people involved to protect their identities. Fighter pilots believe that outside of our community, names and tales don't always mix! And of course, I need to thank my wife, Jacqueline. She came into my life after my time as a fighter pilot and chaser of adventure. As Jacqueline heard the stories from my friends, read the scattered newspaper articles and award citations about my exploits, and listened to my tales, she realized something lost to me; that I have lived a very charmed and storied life. Jacqueline urged me to write about my life and times as a combat pilot, so that our son Alex, and my daughter Katie, would know more about their father. Jacqueline encouraged me through the long hours spent in this endeavor. Thank you,

Jacqueline…you are my heart and my happiness, my best friend, and the love of my life!

Chapter 1: Damn, it's good to be me!

My callsign that day was Luger 31. It was a beautiful, sunny day over central Yugoslavia. I was cruising along at 20,000 feet doing .8 Mach, leading a four-ship of F-16CJs. My wingman, Luger 32, was in tactical formation to my left, about 9,000 feet away and stacked a thousand feet higher than me, his jet shining in the bright sunlight. I scanned the sky from left to right, high, and low, looking for MiGs or any other threats. I turned my attention inside the cockpit for a brief moment, checking the radar for airborne contacts and my other sensors for indications of enemy radars that might target us. Seeing nothing on the fancy equipment, I returned to searching the sky with my eyeballs. I twisted in my seat and checked behind and underneath Luger 32, then looked further back where I could just make out Luger 33 and Luger 34 flying a couple of miles behind me. "Looking good," I muttered to myself in my oxygen mask. Fighter pilots pride themselves on professionalism, and today Luger flight looked sharp and in perfect formation. Flying in a perfect formation isn't just about looking good, it's also important tactically. Our formations are flown to provide the best combination of mutual protection, maneuverability, and lethality. I smiled with pleasure and continued my visual search of the sky and ground around us. People think that with our modern radars and sensors that nothing goes undetected, but the hard reality is that probably 90% of the time when someone gets shot down, they never saw it coming. So, like the guys flying Sopwith Camels in the skies of France, Corsairs over the blue waters of The Slot, Sabres in MiG Alley, or Thunderchiefs along Thud Ridge, F-16 pilots keep their heads on a swivel "checking six." As I continued my visual search, I paused to stare at Obrava Airbase, a Yugoslavian military airfield passing just to my right. I looked at its runway and taxiways and parking ramps, searching for signs of any activity worth noting.

The pair of F-15Es we were supposed to protect that day while they bombed a target had cancelled for some reason. But since we were already there, I decided to lead our formation around over central Yugoslavia as a show of force. Officially it was called "providing presence," and was intended to make sure the enemy didn't forget that we owned the sky over their country.

Apparently, the Yugoslavians weren't impressed or intimidated, and as I stared at the airfield below, I saw it happen; they shot a Surface to Air Missile (SAM) at me. SAM launches have always looked a bit surrealistic to me, and this time was no different. Just as the missile began to lift off, my wingman called "SAM launch 2 O'clock low!" I stared at it, for what seemed like a long time, noting the details…details I had become quite familiar with during dozens of similar attempts to shoot me down. It was an SA-3, which is an older SAM system designed and built by the former Soviet Union. Old or not, it was still deadly effective. Only a few days earlier an SA-3 had shot down an F-16 near Belgrade. Just as the missile came off the rail of its launcher, my Radar Warning Receiver (RWR) began flashing and screaming warning tones into my helmet indicating a missile launch. SA-3s look for all the world like a Saturn V rocket launch; boiling red-orange flames and a donut of dirt blown up as the missile's rocket engine lights-off and the missile body vaults into the sky. My pulse quickened as the ever-so-faint shape of the missile became visible, emerging from the cloud of fire and debris...ascending...accelerating...and then tipping toward me. The SA-3 may be an older missile, but its top speed is actually quite high. It looked slow at first, but quickly picked up speed and maneuvered steadily toward me as it began responding to guidance signals from its parent radar.

My reaction to the launch, while very deliberate, was executed without much thought...the result of muscle memory from hundreds of repetitions in training and combat. "Luger 31, SA-3 launch southeast, defending 060 degrees" was my radio call, letting my flight-mates know I was the one being shot at, what was shooting at me, and which way I was dodging. As I spoke, I pushed the throttle forward to its stop, rotated it outboard, and forward again to select full afterburner. The acceleration pressed me back in my seat. Speed is life, and my jet would need all of the energy I could generate to make sure that I could defeat this SAM shot. Without taking my eyes off the missile, I deployed bundles of metallic chaff designed to decoy radars using the "slap switch" on the cockpit wall next to the throttle. At the same time I activated my Electronic Countermeasures (ECM) pod with another switch on the stick. The F-16 is designed so that almost every function you need to fly, and fight, can be accomplished without taking your hands off the flight controls...a concept called

"HOTAS," which stands for "Hands on Throttle and Stick." Having HOTAS is a big plus when someone is shooting at you, letting you keep your focus outside of the airplane instead of having to look around inside the cockpit to find the right switch.

While designed to confuse radars, ECM and chaff alone aren't guaranteed to prevent a missile from hitting you, but they certainly help. I watched as the missile angled itself on an intercept course with my jet. I crisply checked my aircraft a little to the left to place the missile on my right wingtip. I then rolled my F-16 upside down and began tracking the nose down toward mother earth with a smooth 4-G pull…all the while keeping my eyes locked on the missile. Sometimes when you begin to maneuver you can see right away that the missile is having trouble guiding on you, or has even lost you entirely, drifting away from your flight path. That didn't happen this time and the missile corrected its course and continued its intercept. One of the things that happens in the movies is that it takes a long time for a missile to launch and intercept you. On the big screen, actors are chased around the sky, under bridges, and around mountains while they keep a running dialog going. In reality, you only have a handful of seconds. There is no time for indecision or pithy dialog. As soon as I saw the missile correct its course in response to my maneuver and continue toward my jet, I rolled my F-16 back upright and began another 4-G pull, tracking my nose back above the horizon, and creating another change in the intercept geometry for the missile to solve.

It was a beautiful, clear blue-sky day, which makes Anti-Aircraft Artillery (AAA) fire from guns on the ground difficult to see. As I pulled skyward, I could just make out some tracer fire way below my jet; medium caliber stuff…37mm or 57mm most likely. AAA is a lot easier to identify at night, as the colors and patterns of the tracer rounds are very distinctive. During the day you often can't see the colors and patterns of the tracers well enough to make a definitive identification. This was the case now but based on experience I knew I was probably right. The AAA would have to wait, and I redoubled my focus on the missile.

The missile itself was getting close enough now that I could see some detail on it. The booster-rocket had dropped away so there wasn't a good flame or smoke trail to follow any more. The missile had corrected in response to my latest maneuver and was getting a lot closer. I heard one of my wingmen call out "Magnum," which was the code-word indicating that he had fired a radar seeking missile called a HARM at the SAM site. HARM stands for High-speed Anti-Radiation Missile, and it is indeed a very high-speed weapon that flies at several times the speed of sound. As fast as a HARM is, it would never get to the SA-3's parent radar in time to save me...but it was nice to know that somebody would pay for capping me if I got hit.

The SAM was now very close. "Time's up," I thought to myself, realizing that I would need to fly my jet precisely and aggressively in a last-ditch effort to defeat the SAM. Double-checking that the throttle was still indeed parked in full-afterburner, I rolled the F-16 into the missile, increasing G at the same time. Straining under the high-G maneuver, I watched the missile pass under me as I rolled thru inverted. It was close, but not close enough for the proximity fuse on the warhead to detonate. I snapped my eyes back to the airfield, looking for a second SAM launch. I could tell from the sluggish way my jet finished the maneuver that my airplane's speed and energy were low, so I paused the jet's roll, flying upside down and letting the nose drift toward the ground to pick up speed. At that moment my RWR sounded a new alarm, indicating that another radar was looking at me. This time it was an SA-6 locking its radar onto my jet. Looking around outside the cockpit I couldn't see the new threat, but I couldn't take the chance that there wasn't a missile closing in on my airplane. I keyed the radio again and called, "Luger 31 still defensive...SA-6 launch, close." I was still inverted from defeating the SA-3, and my airspeed was also slower than I would have liked. I was low enough that the AAA I previously considered a mere curiosity would soon constitute a very real threat if I didn't pay attention to it. And if the SA-6 really had locked on to me and launched, I was in the heart of the missile's lethality envelope. "Fuck this!" I said into my oxygen mask, and since I was already upside-down I hauled back on the stick. My F-16 responded like a true thoroughbred, and the nose whipped straight down into the vertical. I was now pointed straight at the ground, passing 15,000 feet in full afterburner. I knew that I would soon over-speed my external fuel tanks and weapons, and most likely over-G the jet when I pulled up to avoid the ground, but it really didn't matter to me right then. The airspeed and G-limits for the airplane are really only important if you think the aircraft will have another chance to fly...something I wasn't sure about at that moment! I figured that in broad daylight, with AAA and SAMs all around me, my safest bet was to get low and fast.

The ground was rising up at incredible speed, and I was starting to sense a little ground-rush from the approaching trees. I judged the speed of my jet (supersonic) and the altitude, and yanked the stick full aft, slapping on a spine-crunching 9-G pullout to just miss the ground. My timing was sweet, and I eased off the G, leveling my F-16 out just above the treetops headed away from all Obrava Airbase. My F-16 rocketed across the ground in full afterburner. Down low, as close to the trees and houses as I could stand, the air was thick, and the GE-129 engine was pumping out a healthy 29,000 pounds of thrust. That's more horsepower than the entire starting lineup at the Indy 500!

Flying supersonic at high altitude is somewhat anti-climactic since you don't have anything around you with which to sense the speed. Flying across the treetops supersonic is totally different, and the sensation of speed is all consuming! My attention was focused on what was directly in front of me, and I twisted the F-16 left and right to throw off the aim of any gunners who might see me coming. Skimming the treetops at Mach 1.2, my peripheral vision registered a green blur, and objects forming in front of me grew from distant specks to flash under my feet in seconds. I wanted to stay low and fast until I was away from the threat area, and that wouldn't take too long. Flying at Mach 1.2 on the deck is going about 900 miles per hour, or 15 miles per minute, which is one mile every four seconds! My pulse was racing, and gallons of gourmet-quality adrenaline coursed through my body. I didn't want to climb back up to altitude…not because of the SAMs and AAA…but because the rush of flying that low and fast was that amazing. It was exquisitely intoxicating! Sadly, it also uses a LOT of gas. The fuel flow meter was pegged at 99,999 pounds per hour consumption, meaning I could only fly like this for a short time before I would run out of fuel.

As soon as I felt sure that I was away from the threat, I smoothly pulled the stick back until the jet was climbing vertically. As I rocketed away from the earth I twisted around and looked back, checking for AAA and SAMs, while pumping out flares to defeat any heat seeking man-portable SAMs that might be around. It was a breathtaking sight. At this speed my rate of climb was about 50,000 feet per minute going straight up. The view over my shoulder was like the one you see from those cameras mounted on the Space Shuttle looking backwards toward the launch pad during blast-off. The ground was dropping away at an astonishing rate, the details of the buildings and trees shrinking, all framed by the small fingers of smoke from my flares. I don't have the words to describe how artistically beautiful it was. Like so many things I have seen and done in an airplane, that exact moment will live in my heart and mind's-eye forever, but no words will ever fully convey the experience.

Passing 20,000 feet I added G once again and pulled the nose of my F-16 back to level flight, rolled upright, and checked the throttle out of afterburner. My RWR was silent, and nobody was shooting at me now. I took a deep breath and was starting to think about rejoining with my flight when the Airborne Warning and Control System (the big, four-engine jet with a frisbee-shaped radar on top of it, also called an "AWACS" for short), callsign Magic, made an urgent broadcast over the radio. "Magic, picture, single group bullseye 168 degrees for 42 miles, 25 thousand feet altitude, tracking south…Outlaw." "SHIT HOT!" I thought. A "bullseye" is a designated place on the ground used as a common reference to describe where something is located. Magic had just given the location of an airplane and identified that it as an "Outlaw," meaning it had originated in enemy territory. "Maybe I will get a MiG kill today?!?!" I thought to myself. In that same second, I realized that Magic had just given the exact bullseye location, altitude, and direction of travel as *my* jet. In a flash I knew what had happened. The AWACS had lost track of my aircraft as I violently outmaneuvered the missile and descended to low altitude, and when I climbed away from the ground so explosively the AWACS mistook me for an enemy fighter taking off, perhaps from Obrava Airbase. Just as quickly my "SHIT HOT!" turned into an "OH SHIT!" as my RWR scope came alive with indications that a Navy F-14 Tomcat had locked its fire control radar onto my F-16. A fraction of a second later at least two F/A-18s locked on to me as well. This was really bad, because unless the Navy pilots where being very disciplined in completing all of the identification requirements listed in the Rules of Engagement (ROE), I probably had friendly missiles already launched at me! Navy pilots are brave to land on that boat at night in bad weather to be sure, but at that moment I wasn't willing to place my life in the hands of a service whose traditions were founded on rum and the lash. I had to act quickly.

"Negative, Negative. Magic...bullseye 168 for 42, 25 thousand is a friendly...Luger 31," I called out over the radio, trying to keep my voice calm. Scanning the sky around me, I expected to see the contrail of a friendly-fire missile headed my way at any second. I began to weave my airplane and put out more chaff to complicate the intercept of any missiles headed my way that I didn't see. Magic didn't reply, but the radar spikes from the F-14 and F/A-18s disappeared. I started to calm down a bit and realized how thirsty I was. My throat was dry, and I was dripping with sweat. Large doses of high-quality adrenaline will do that to you.

After a few seconds of blissful silence from the RWR and radio, I figured I was safe enough, and turned north to rejoin with my flight...giving Obrava Airbase a wide berth! In short-order I located my flight mates, and we were soon assembled in a perfect tactical formation once again. That accomplished, we were out of gas, and it was time to head back to Aviano Air Base, in northern Italy. The sun was dipping lower in the sky as we flew out from bad guy land and headed to the aerial refueling tanker to get some gas. The flight back to base would take us about an hour, and we needed to have enough fuel in case our landings were delayed. As soon as we were out of Yugoslavian airspace, I pulled out a water bottle, unlatched my oxygen mask, and emptied most of the delicious contents in one long, gulping, pull. Lowering the bottle, I looked around at the mountains of Albania sliding below. They still had snow on their peaks and were ruggedly beautiful. Looking ahead, the Adriatic Sea was just visible in the sunlit haze, with the low sun reflecting off the water in a slowly changing pattern of sparkles. Arranged around me, my flight mates were still flying in perfect tactical formation. I twisted around in my seat and, looking out from the cockpit, gave my F-16 a good look-over. I thought to myself what a marvelous jet she was, and I was one lucky guy to get to ride her every day! I drained the last of the water and stowed the empty bottle. I imagine that my wingmen watched in amusement as I pulled the F-16's nose up about 15 degrees and pushed the stick to the left, snapping off three aileron rolls, smiling like a fool as I watched the world roll around and around. "Damn," I thought, "it's good to be me!"

Chapter 2: The only thing I ever wanted to be was a fighter pilot

I always wanted to be a fighter pilot. I can't remember ever wanting to be anything else. In fact, my family can't remember me ever wanting to be anything else either. They all say that from the time I could talk, which was annoyingly early, I wanted to be a fighter pilot. I never talked about being a doctor or a lawyer or a cowboy or a bank robber...always a fighter pilot. My earliest memories are of pretending to fly airplanes. It was all I ever talked about, and I am sure I must have bored my best friend, Jeff Thomas, to death over the years with my musings on the subject. I must have built a thousand model airplanes growing up, saving every nickel so that I could buy a new Corsair or Thunderbolt kit. As soon as the glue and paint dried (and sometimes before), I would run around the front yard making airplane sounds, strafing imaginary targets, and fighting epic dogfights with Japanese Zeros and German BF-109s. When I wasn't building model airplanes, I was reading every book I could find about flying. I don't remember what other kids were reading, but I always had my nose stuck in a biography or autobiography of some fighter pilot from the past. I read books by friends and foes alike, including Robert Scott, Eric Hartmann, Pappy Boyington, Saburo Sakai, Douglas Baker, Stanford Tuck and many others. I dreamed of one day having great adventures and flying fighters in combat just like they did.

My Mom and Dad always encouraged me too, although I'm sure they saw my fascination as the fanciful dreams of a young boy. We were a pretty poor family. I have a brother and two sisters, but they are a lot older than me. In fact, two of them were already married when I was born. I was a post-menopause surprise for Mom and Dad, and I used to imagine what they must have thought when they found out I was on the way at their age. I eventually found out, when Jacqueline and I got pregnant and had our son Alex in our 50's! My dad only had a third-grade education, but he had an amazing work ethic. Pop labored like an animal every day of his life, and Mom worked long hours at a local daycare. As a result, I always had a roof over my head, clothes on my back, and something to eat. But I can't over-exaggerate it; we were poor. "How poor," you might ask? During the winter my dad would take the Styrofoam tray that the hamburger came in, wash it, and trace the shape of my foot on it. He would then cut it out and place it in my shoes to help keep my feet warm. That kind of poor. In spite of the fact that we had nothing, my folks were true believers in the American Dream. I was taught at an early age, that if you were willing to work hard enough, you could do anything or be anything. So, mom and dad encouraged me to become a fighter pilot if that's what I wanted to be. They put up with the smell of model airplane glue and hauled me to airshows whenever they could. Dad never lived long enough to see me become a pilot, but mom helped pin on my Air Force pilot wings one day at Vance Air Force Base, in Oklahoma. I'm living proof that the American Dream is real!

One day when I was 12 years old, I opened up the local newspaper and there was an article about the Warbirds of America organization. They were, and still are, a group dedicated to restoring and flying former military airplanes. I was enthralled at the idea of getting to be around the airplanes I saw at airshows and had read stories about. The article interviewed a guy named Jeff Ethell, who was, as I recall, the head of the local Warbirds of America chapter. He was a much-published author on warbird aviation, and actually owned his own airplane...an SNJ-5 Navy training plane from WWII. I wrote Mr. Ethell a letter, care of the newspaper that printed the article, explaining my love of airplanes and my dream of one day becoming a fighter pilot. I asked Mr. Ethell if there was a "Junior Warbirds" organization for kids like me, and if so, how could I join. Mr. Ethell actually wrote me back. As I read his letter, I was very disappointed to discover that there wasn't any kind of "Junior Warbirds" association I could join. But my disappointment quickly turned to excitement when I read his invitation for me to come out and fly in his SNJ if my folks would let me do it!

On the appointed day my dad drove me out to the local airport where Mr. Ethell kept his airplane. I remember every moment of that experience. As the hangar doors opened, I stood in awe looking at the huge, blue beast in front of me. SNJs really aren't that big of an airplane but it seemed enormous to me at the time, and in my mind's-eye I can still see the radial engine and propeller shining in the sunlight! My dad helped Mr. Ethell push the airplane out of the hangar, and then Mr. Ethell settled me into the back seat and strapped me in. With a smile and a clap on the shoulder he slid the canopy shut over me and crawled into the front cockpit. While he was strapping himself in, I sat there soaking it all up…the myriad of dials and levers and switches, the stick and throttle worn free of paint from a thousand hands on the controls…and the smell. Ahhh, the smell! It was a mixture of aviation gas, and oil, and hydraulic fluid, and zinc chromate, and leather and sweat. I didn't know it at the time, but I would smell pretty much that same smell in every military airplane I would ever fly…from 1930's biplanes to F-16s. As I write this, I can close my eyes and inhale through my nose, and the memory of that smell is as real and present as if I was sitting in an airplane right now. I think it's the most wonderful smell in the whole world!

Mr. Ethell checked in on the intercom, and before I knew it the big radial engine rumbled to life, and we were taxing out. An SNJ is not a very fast airplane by any measure, but at the time, that takeoff seemed like I was being shot out of a cannon! As the airplane accelerated the tail lifted up, and soon we were airborne. We climbed out of the airport and Mr. Ethell kept up a running commentary about the airplane and what he was doing to fly it. I stared out of the canopy in awe, and I remember thinking how loud it was, between the radial engine and the sound of the wind rushing over the airplane. I was totally lost in the experience, overwhelmed with the sights, and sounds, and smells of being in a real airplane! I hadn't said much up to this point, but I quickly replied, "Yes Sir," when Mr. Ethell asked me if I would like to see a loop and a roll! My heart pounded. Not only was I getting a ride in this airplane, but I would also get to experience some of the maneuvers I had seen at so many airshows! Mr. Ethell raised the nose of the airplane up, and I watched in amazement as the ground and sky rolled around in a series of aileron rolls. After the last roll he let the nose of the airplane drop, and the wind noise grew as the airplane accelerated towards the ground. I saw the stick move back, and for the first time I experienced the sensation of G-forces as the airplane climbed into a beautiful loop. Up, up the nose went, and as the airplane floated across the top of the loop, I looked straight up out of the top of the canopy and was transfixed by the sight of the world upside down above me. It was amazing to behold! As the airplane accelerated down the backside of the loop, I once again felt the G-forces increase, and then we were back to straight and level flight. Mr. Ethell asked if I was okay. Today I smile and imagine that he was worried he would have to clean up kid-puke in the rear cockpit! Again, I answered with an enthusiastic "Yes Sir!"

I was more than okay, I awestruck! Before I knew it, the landing gear thumped down, and we landed and taxied back to the hangar. I can still remember my dad standing there smiling as we pulled up. As the engine shuddered to a stop, and Mr. Ethell helped me out of the rear cockpit, I wanted to be a fighter pilot more than ever!

Interestingly, that wasn't the last time I ever saw Mr. Ethell. Many years later, just after returning from the first Gulf War, I was at the Oshkosh Airshow in Wisconsin. Oshkosh is the most amazing annual event in all of aviation, and I was fortunate to be stationed nearby and have the chance to attend. It was a hot and humid day in early August, and I was strolling up and down the rows of warbirds (always my favorite kind of airplanes) when I stopped to look at a beautiful Supermarine Spitfire. And there was Mr. Ethell sitting in front of it in a lawn chair! I walked up to him, and he squinted up at me from under the bill of his baseball cap. "Mr. Ethell," I said, "there's no chance you remember me, but my name is Tom Littleton, and when I was a boy, you took me for a flight in your SNJ!" He looked at me for a second and stood up. His face broke into a very warm smile, and he shook my hand with genuine enthusiasm. "I do remember you!" he replied. We stood there talking for quite some time, and I shared with him that I was in the Air Force flying jets, and that the kindness he showed in taking a young boy flying was never forgotten. It turned out he had sold the SNJ some years after I flew with him, and that now in addition to writing articles and books, he spent a lot of time flying rich people's warbirds to airshows...hence the Spitfire he was sitting next to. After we chatted for a while, I shook his hand and told him I would leave him be. But before Mr. Ethell let go of my hand, he asked if I was available for dinner that night. I of course said yes, and we arranged where to meet later.

You can imagine my surprise when I sat down to dinner that evening and discovered our table mates included a collection of WWII fighter pilots, including Eric Hartmann, the Ace of Aces! That dinner was as amazing to me as my first flight in that old SNJ so many years ago. Talking with Mr. Hartmann (I speak pretty good German and he spoke wonderful English) was a once in a lifetime experience that I will always treasure. And while the old pilots indulged my questions and told me their stories, I was shocked that they all wanted to hear my tales of flying modern jets, and what my experiences in Desert Storm were like. I think that they all wished they could be me!

Years later I happened upon a news article about a WWII P-38 Lightning that had crashed, killing the pilot. I was saddened to learn that Mr. Ethell was the pilot killed that day. I sat back in my chair and reflected on the impact Mr. Ethell had on my life, and the inspiration he provided to a young boy who dreamed of conquering the sky. Mr. Ethell, thank you!

Many years passed from the SNJ-5 flight before I found myself in the cockpit of an airplane again, and Mr. Ethell wasn't the only person to have a lasting impact on my quest to fly fighter planes. My dad died while I was a young teenager, and if my family was poor before that, it was nothing compared to life after my Pop's passing. I don't know how my mom managed to get us by, but she did. Needless to say, there was no way I could afford to learn to fly when it was tough to buy food. I graduated from high school and was eventually accepted to Randolph-Macon College in Ashland, Virginia. I was a horrible high school student and barely graduated, although I had exceptionally high SAT scores. I applied to several colleges, but my SAT scores alone weren't enough, and I didn't get accepted anywhere. Fortunately, the Methodist Pastor at our church, Reverend Dr. Riddick, took special notice of me…why I will never know. When he found out that I didn't get accepted into any colleges, he took it upon himself to write a letter to Randolph-Macon College, which is a Methodist-affiliated school. I never saw what he wrote in that letter, but he convinced the admissions folks at Randolph-Macon that I was worth taking a chance on.

By the end of my freshman year in college he might have been second-guessing his efforts; I was more interested in chasing women and drinking beer than going to class or studying. I finished the year with a GPA of 2.01, which was only .01 above being placed on academic probation. But then a funny thing happened. That summer I was working in a department store selling women's shoes, and a few doors down was a U.S. Air Force recruiting office. One day, while on a break, I decided to drop by the Air Force recruiter and see about becoming an Air Force pilot. The recruiter explained that to get into the Air Force through Officer Training School and become a pilot I needed a college degree. It didn't matter in what subject I majored, as long as I achieved a minimum of a 3.0 GPA. I would also need a competitive score on the Air Force Officer Qualifying Test. He added that a private pilot's license would look good on my application if I could get one. And of course, I would have to be medically qualified, although I wasn't worried about that at all. I walked out of the recruiter's office and did some serious soul-searching over the days that followed. I realized that my dream of being a fighter pilot was doable, but I would need to get my act together and focus on school. I would need to save some money and learn to fly as well. Before I went to sleep one night, I made my mind up. Mom and dad were right, in America you can be anything you want to be if you are willing to work for it. Somebody gets to be a fighter pilot, so why not me? And my life was never the same after that moment. I had found my motivation. I quit going to parties and chasing women…well, at least I slowed-down a lot! I earned straight "A's" from there out. I worked three part-time jobs to pay my bills and tuition, and to save some money for flying lessons. And at the end of the next school year, I walked into the "ZZZ Flying Ranch" flight school at Hanover County Airport, Virginia, and told them I wanted to learn to fly.

The ZZZ Flying Ranch was an interesting place, owned and operated by a former Air Force fighter pilot who went by the nickname Fuzzo, because of his high-and-tight haircut. Fuzzo was a real character. I remember him driving around on an old motorcycle with a sidecar attached to it and being grumpy all the time. He was also something of a local celebrity, doing live traffic reports from the cockpit of a Cessna-152 for a local radio station each morning. There were a lot of great flight instructors at the school too, but in a wonderful stroke of luck I was assigned to Douglas Bird.

Doug was an old Englishman, and over the coming months I found out that he was something of a legend, and absolutely the perfect instructor for me. Doug was a short man, a little rotund, and his balding cranium was ringed with a fringe of white hair that always seemed slightly disturbed like someone had just ruffled it with their hand. He walked around with his hands clasped behind his back, slightly bent forward, one eye squinted shut a bit, and looking at you side-long with a knowing grin. He was polite to the extreme and spoke very gentlemanly, saying things like "Quite right, old chap" or "Indubitably, Sir." In total, the effect was amazing…he was the very caricature of an English gentleman from some remote corner of the British Empire of old. Which is exactly what he was!

Doug grew up in Kenya between World War I and World War II. His father was a large landowner and farmer there. Doug and his father became interested in flying in the early 1930's, and mail-ordered a DeHavilland Tiger Moth biplane from England, which they assembled in a barn. They then proceeded to teach themselves how to fly it! As World War II loomed on the European continent, Doug travelled to England and joined the RAF. He was a Spitfire ace and flew in the Battle of Britain. Later in the war he was transferred to fly bombers, and one night his plane was shot down over France. As Doug prepared to bail out, he discovered that his bombardier was pinned in the airplane by the damage and unable to get free. Rather than leave that man behind, Doug climbed back into the cockpit. Unable to jettison the bomb load because of the damage, Doug made the decision to stay with the airplane anyway. He safely crash-landed it in the dark, in a French farm-field, saving his bombardier. After the war, Doug was awarded the Victoria's Cross, England's highest decoration, for his valor that night. Doug was captured by the Germans and made a prisoner of war…but not for long. He was a part of "The Great Escape" of POWs, and along with five other escaped Allied pilots, strolled onto a German airfield at dawn one day and stole a JU-88 bomber which they flew back to England and freedom. After the war, Doug returned to Kenya and ran a bush-flying operation until Kenya became independent from the British Empire. Doug then returned to England and became a British Airways pilot. And by the time I climbed into a Cessna 152 with him for the first time, he had retired as a

Boeing 747 Captain with more than 45,000 total hours of flying time!

Doug taught me more than just how to fly…he taught me how to think about flying. One day early in my training, he was following me around the airplane as I was doing the pre-flight. As I finished my walk around, I noticed he was standing out in front of the airplane about 15 feet away, hands clasped behind his back, looking very philosophical. I walked over to him, and he cocked his head sideways and looked at me. "Tom, never be complacent about flying. Never climb aboard without knowing you are ready for whatever may happen. The day you look at an airplane and don't feel respect for what can go wrong, it is the day you should stop flying." I nodded sagely, but of course I didn't really understand what he meant. However, I never forgot what he said, and I thought about it every time I walked around an airplane…be it a Cessna or an F-16. Eventually I learned the wisdom of his words. Flying demands respect and 100% of your attention. It is true that you never know what will happen when you slip the surly bonds of earth and fondle the face of God (read the poem, "High Flight" to get my irreverent reference!). You have to know that you did everything possible to prepare for the flight, from planning, to thinking through what you will do if something unplanned or unexpected happens. The last thing I would always do before climbing into my jet was to walk out in front of it and take a knee. As I would stare at the jet, I would ask myself Doug's question. "Are you ready for this?" The answer was always a resounding, "Yes!" and that's a big reason I have lived to be both a bold pilot and an old pilot.

Doug also taught me another big lesson about flying. One day we were doing steep turns, and I was having a devil of a time maintaining a steady altitude and bank angle. When I was done hauling the airplane around like a drunken porpoise, Doug cocked his head sideways and gave me that side-long squint that foreshadowed a lecture. "You know, old chap," he began, "airplanes are better at flying than we pilots can ever be. It will do anything you want, and beautifully too, if you just ask it properly." He then continued on to explain that if you apply the right combination of power, flight control movements and trim (aerodynamically balancing the controls), the airplane will fly itself smoothly through whatever maneuver you want. He also explained that if you do that all the time, and the airplane is not cooperating, you should think twice about why. He stressed to never let the airplane fly me, but to always give the airplane a chance to fly *for* me. It changed the way I flew, and years later when teaching new pilots to fly the F-16, it was a lesson I always passed along.

Sadly, about a year after teaching me to fly, Doug was killed in a Cessna 172 crash. He was the first of several dozen friends I would lose over the years in airplane crashes. I will always think of him as the Obi-Wan Kenobi to my Luke Skywalker.

Another man who had a lasting and profound impact on my flying was John Greenwood. John was a former Air Force pilot who had fought in the Korean War. John owned a Pitts S-2A…a hotrod little biplane that he flew in airshows all over the country. I took aerobatic lessons from John after I earned my private pilot's license. John was outspoken, opinionated, tall, tan, fit and Hollywood-handsome…the polar opposite of Doug. But like Doug, he was an extraordinary pilot and demanded perfection in the airplane. He would never let me get away with the slightest mistake, and if I lost a single foot of altitude during a slow-roll he would make me do a dozen more to get it right. His attitude about flying dovetailed with what Doug had taught me…never let the airplane fly you! John used to roll his Pitts out in front of his hangar, and then pull out a couple of lawn chairs. He would spend hours chatting with different pilots, myself included, telling his flying stories. Those are some of the fondest memories I have from my early days of flying, and I remember wishing that I could have adventures like the ones John and Doug talked about. Years later my brother Roy and his wife Linda, who are both excellent pilots, ended up keeping their airplane in the hangar next to John's. One time I was home on leave from the Air Force, and I went with my brother to see his airplane. John was sitting outside by his Pitts, holding court with a small circle of local pilots. When I walked up, John jumped up and pumped my hand in greeting. Flashing his Hollywood smile, John introduced me to the group as one of his former students. As I sat down in one of the worn lawn chairs, John said, "Tom here is a fighter pilot, flies F-16s, fought all over the world. He could probably teach me a thing or two now." I smiled on the inside as the pilots around the circle started asking me questions and listening with rapt attention to *my* stories.

Chapter 3: Not my first rodeo

The air war in Kosovo wasn't my first-time seeing action. I don't know why, but I was always the guy who seemed to be involved in whatever was happening around the world. So many of the men and women I flew with over my career never had a chance to fire a shot in anger or have a bad guy make a concerted effort to kill them. But not me; I was always where the shooting was. As a young Second Lieutenant, I flew Cold War missions along the Soviet Coast off the Kamchatka Peninsula and across the North Pole to the East Siberian Sea. The night before Saddam Hussein invaded Kuwait I was in Alaska, having just returned from one of those missions. We had been intercepted by a pair of Soviet SU-22s and decided to go out celebrating. Nobody shot at anybody, but it was still a big deal to go beak-to-beak with a Soviet fighter. I felt like Tom Cruise in the opening scene of Top Gun! Me and the guys I was flying with to drive up to a place called "The

Howling Dog Saloon," which we were told was the world's farthest north biker bar. We had a great time, and pushed it up pretty hard, drinking beer, playing volleyball, eating sloppy joes, and arguing with the hippie burnouts who live off the grid way up there in rural Alaska. The next morning, I was laying on the couch in my room at the Bachelor Officer's Quarters at Eielson Air Force Base, hungover as all-hell, when CNN announced that Saddam Hussein had invaded Kuwait. I thought to myself, "I bet this will be a Trivial Pursuit question one day, so I better remember it." Ha…no worries there! About five minutes later the phone rang. Before I knew it, I was deployed to Saudi Arabia for Operation Desert Shield, and on January 17th, 1991, I flew on the opening night of Desert Storm. When we assembled for the mass-brief before takeoff, the Brigadier General sent to give us a pep-talk told us that they expected 20% of us to get shot down that night. Sobering, to say the least! Desert Storm was my first taste of real combat, and I was hooked.

I also flew in Operation Phoenix Palm, which was the pre-cursor operation to full-blown involvement in Somalia. The No-Fly Zones over Southern and Northern Iraq (which I will write about later in this book), the Balkans wars…I got a piece of them all. But when I arrived at Shaw AFB,

South Carolina, in 1996, things really seemed to kick into high gear. On the day I arrived at Shaw I tossed my flight gear into my pickup truck and drove over to my new unit, the 78th Fighter Squadron: The Bushmasters. As I walked up to the squadron, I noticed a Lieutenant Colonel standing outside watching me approach. I saluted, and the Squadron Commander of the Bushmasters returned it and shook my hand. I told him I was his newest pilot, and he asked me "Are you ready to go to war?" I figured this was just his being hardcore, and of course I replied, "yes Sir!" My smile turned to disbelief when he continued, "Well good. Drop your gear and go sit in the main briefing room, we're going to war as soon as the Wing Commander gets here." It turns out I arrived just hours after Cuban Air Force MiG-29s shot down two Cessna-337's flying off their coast on a humanitarian mission. We were going to fly Combat Air Patrols (CAPs) providing Force Protection for the Search and Rescue (SAR) operations that were about to commence! While the mission was cancelled before we could take off, it was sure an exciting start to my time in the 78th Fighter Squadron, and a harbinger of things to come! In the coming months and years as Bushmaster I would spend a lot of time at war, flying missions to enforce the No-Fly zones over Iraq, and eventually in the air war over Kosovo. After my tour in the 78th was over, my combat opportunities kept coming, with deployments to Iraq for the second Gulf War as well as Afghanistan. After spending a career flying combat missions, I was addicted to the challenge and excitement. When I retired from the Air Force, I missed it so badly, that I took a job flying small propeller airplanes as a private military contractor (perhaps "mercenary" would be a more accurate term, but that's not politically correct these days). The work involved flying high-risk security missions in Central and South America, Africa, the Balkans, and back in Iraq and Afghanistan too. That work was far more dangerous than flying an F-16, but when you are an adrenaline junky hooked on flying combat missions it was a risk worth the reward.

There are a lot of good flying stories in between, and I will recount some of them later in this book. The end result of all this combat action was that I quickly became an F-16 instructor pilot in the 78th, and just about a year before Operation Allied Force kicked off, I was made a Flight Commander. A Flight commander in a fighter squadron is the senior pilot in charge of about a dozen other pilots, responsible for supervising them, training them, and leading them in combat. In the spring of 1999, I was at the top of my game as fighter pilot. I was highly trained and experienced, supremely confident, professional, and lethally talented. There wasn't anything that I couldn't do with an F-16, and no one could outfly me. When the word came that we were deploying to Aviano Air Base in Italy for the Kosovo air campaign against the Yugoslavians, I just nodded my head and smiled. "Shit," I thought, "I feel sorry for those ethnic-cleansing Serb bastards…they don't stand a chance against *me*!

Chapter 4: It was like old home week

My first combat mission of Operation Allied Force, the air war over Kosovo, was off to a pretty shitty start. Another instructor pilot and I were the first two Bushmaster pilots scheduled to fly combat missions after the squadron had arrived in Italy. The plan was to fly with two pilots from Spangdahlem's F-16CJ squadron to see how they were running operations. That way we could explain to the pilots in the 78th Fighter Squadron how things were working over Kosovo, ensuring that our squadron's tactics dovetailed with everybody else's. I woke up early that morning, pulled on my flight suit, and was ready to drive from our hotel to the base on time. The other guy I was flying with was running late, however, because he overslept. As a result, we didn't have time to swing by the chow-hall and get breakfast. So, I was starving by the time we got to the jets, and that pissed me off.

My callsign that day was Colt 43, and my wingman was Colt 44. I can't recall the names of the two Spangdahlem pilots flying as Colt 41 and 42, but they gave us a good briefing for our first "KEZ" sortie. "KEZ" stood for the "Kosovo Engagement Zone." The ROE broke the Yugoslavia up into three fairly artificial chunks. The northern section that was basically the northern third of the country and included the capital city of Belgrade and the large city of Novi Sad. There was the middle third of the county that included the large Yugoslavian air base at Obrava. The final third was the southern part that included Kosovo, or the KEZ. The KEZ itself was so busy that it was further subdivided into an eastern half nicknamed the NBA, and a western half that was called the NFL. On that day we were flying a KEZ mission in the NFL. To get there we would fly south down the Adriatic coast of Italy, and then cross the sea to Albania, where we would meet up with a tanker. After refueling we were scheduled to provide protection from SAMs and MiGs to several strike missions. We would be working primarily with an F-16 Forward Air Controller, using the callsign Bear 81. Bear 81 would locate and identify targets, and then control flights of fighter planes armed with various missiles and bombs to strike those targets. We would cover Bear 81 and his strikers in case enemy fighters showed up, or the Serbs decided to shoot SAMs or AAA at them.

As if starving to death wasn't bad enough, my jet broke shortly after I started taxiing out for takeoff. The flight control computer malfunctioned, which is a big deal in the F-16 since the computer actually flies the jet based on what you control inputs you command. Colt 41 and 42 went ahead and took off, and my wingman waited for me at the end of the runway while I went back to the parking ramp for a second jet. I parked and quickly shut down my broken jet, climbed out of the cockpit, and ran to the spare F-16. Now I was hungry, sweaty, and running late, which made me royally pissed off by the time I got to the end of the runway and received takeoff clearance. As I taxied into position, I performed my before-takeoff checks; I stirred the stick to check that my flight controls were free and clear, turned the pitot-heat on, armed the ejection seat, confirmed that no airplanes were on final approach and about to land, tightened my straps, activated the air traffic control transponder, and switched on the landing light and aircraft camera system.

I lined up on the runway and watched as my wingman pulled into position beside me. He braked to a stop and gave a big head-nod; the signal that he was ready to go. Fighter pilots use a lot of hand and head signals to communicate without using the radio, and I responded to his head-nod by twirling my right forefinger in circles beside my head to have him run-up his engine. I did the same, and after checking my own engine instruments, I looked over at my flight mate. His run-up was complete, and he gave me another big head-nod indicating that his engine was good, and he was ready to go. I gave him a thumbs-up and a quick salute, released the brakes, and moved the throttle into full-afterburner. The fully loaded F-16 rocketed down the runway despite carrying a full combat load of weapons: 510 rounds of 20mm ammunition, an ECM pod, buckets of chaff and flares, two 370-gallon external fuel tanks, two AGM-88 HARMs, two AIM-120 radar-guided Advanced Medium Range Air-Air Missiles (AMRAAMs) and two AIM-9M Sidewinder heat-seeking air-air missiles. Passing 90 knots the rudder became fully effective and I disengaged the nosewheel steering. In a heartbeat I was accelerating through 190 knots and the jet flew cleanly away from the ground with only a small aft movement of the stick. I raised the landing gear and climbed out of Aviano at 350 knots. Soon, my wingman slid his jet into tight formation on my right wing. I looked his jet over closely to make sure nothing had fallen off during takeoff or that any panels had come loose. Staring at his jet, loaded with the same weapons as mine, I smiled and thought to myself, "God, we have a cool job!"

At that very moment a bunch of caution lights went off in the cockpit as my inertial navigation system (INS) failed. "Damn it!" I muttered in my oxygen mask. The INS is not only important for navigation, but it also happens to provide important data to many of the systems in the F-16 and is a key component of accurate weapons delivery. Without an INS you can still fly and fight if you have to, but you would never take an F-16 into battle without the INS if you could avoid it. I wasn't too concerned since INS's do fail sometimes and you can usually re-align them inflight.

The sky was absolutely clear, which made it easy for me to navigate while I worked on my problem…all I had to do was look outside and fly south down the coastline. So, I set about running the checklist to bring the INS back up. Before long I had the system working again, and just as I began to think about catching up to Colt 41 and 42, the INS failed again. So, I performed another in-flight alignment.

And it failed again. And again. Six times in total. There was no way I was going to abort my first combat mission over Kosovo. People would think I was a pussy and scared to fly. I made up my mind that I was going to complete my mission regardless of having the INS or not. I finally got the INS to work and stay working on the seventh try, but by then I was very frustrated…on top of being hungry, sweaty, and late! By the time I was flying a fully combat ready jet again we were almost to the tanker over Albania.

The tanker that day was a KC-135E, callsign Guilder 60. If the weather is bad, and there are a lot of airplanes going to and from the refueling track, it can be a lot of work to find the right tanker and get rejoined. But it was a beautiful day over Albania, and the rendezvous with Guilder 60 went smoothly. As we approached the tanker from behind and below, I opened my air-refueling receptacle and watched my wingman move ahead to fly formation on the tanker's wing until it was his turn to get gas. I slid slowly into position below and behind the tanker and was soon hooked up and refueling. The big stick that lowers from the tanker and hooks up to give you gas is called "the boom." When the boom is connected to your jet, it also connects you to an intercom system in the tanker so you can talk to the crew without having to transmit over the radio.

While I was taking gas, I was chatting with the crew, and I was pleasantly surprised to discover that the Aircraft Commander of the tanker was an old friend of mine, a guy named Jay! We used to fly together at another time and place in the Air Force, and we had some crazy adventures together involving airplanes, wine, women, and song…the kind of shit you'd be embarrassed about if your mom ever found out! Before we could do much more than say hello, my gas tanks were full. It was my turn to fly on the tanker's wing while my wingman, Colt 44, got a top-off. Sitting there in formation on the tanker I could see Jay in the pilot's seat, prematurely bald, and grinning at me from behind his Oakley sunglasses. He nodded his head up and down like he was saying "Alright!" as he looked over my jet. I raised up the sun visor on my helmet and dropped my oxygen mask to the side so he could see me smile in return. As we cruised along in silence, I thought, and not for the last time that day, what a wonderful brotherhood I was a part of. As I was reminiscing about old times together, I realized Jay had a camera and was taking my picture. I made a mental note to call his squadron after my mission, chat with him some more, and give him my address so he could send me a couple of those photos.

Before long Colt 44 had his gas and it was time to go. We were already late for the war, and I didn't want to miss anything. Jay looked over at me and waved. I reconnected my oxygen mask, dropped my sun visor over my eyes, and held up my middle finger, flipping Jay off! He busted out laughing and waved again as I slowly began a climbing turn away from his jet.

It was only a few minutes flying time from the tanker track to the edge of Yugoslavia's airspace. As we left the tanker, I repositioned my wingman so that he was flying in tactical formation off my left wing; about a mile and a half line-abreast and a thousand feet higher in altitude. This formation allows both flight members to see behind and below their wingman, checking for enemy fighters, SAMs, and AAA. It's also far enough apart that you can look at your sensors and scan the sky around you, and the ground below you, without too much chance of accidentally running into each other. Being close enough to see your wingman's airplane is important, a concept we call providing visual mutual support. After confirming my wingman was in position I keyed the microphone, making a short radio call directing him to change his main radio to the RC-135's frequency. The RC-135 is a four-engine jet, and it is loaded with electronic warfare equipment. I contacted the crew of the RC-135, and they provided me with an update on all of the enemy's radar activity, which I jotted down on my kneeboard. Next, we switched over to the strike frequency. I changed my radio and I listened to the chatter for a few seconds to build my situational awareness.

Over our private inter-flight radio, I directed my wingman to "Fence-In." When you approach bad-guy land you do what is called a "Fence Check," or "Fencing-In." This prepares your airplane for battle. When you leave bad-guy land you "Fence-Out" to safe your weapons and turn the F-16 back into a "normal" airplane again. While I digested what I heard on the strike frequency, my hands moved around the cockpit performing the long-practiced ritual of Fencing In. I started by turning off my exterior navigation lights, then flipped a toggle switch that depressurized my fuel tanks and pumped the dead-space inside of each with Halon 1211, which is a fire suppressant. The Halon would prevent any fuel vapors in the tanks from exploding if I took a hit. Next, I checked that my ECM pod was correctly set up and functioning, before placing it into a standby mode so it would be ready if I needed it. I also armed my flares and chaff, expending one of each to make sure they were functioning properly. I turned up the volume on my RWR, and then quickly ran a check of all my weapons displays, sensors, and air-to-air radar to make sure everything was set the way I liked it. I then made a last sweep around the cockpit stowing any loose items. Finally, I reached over and moved the Master-Arm switch from "Safe" to "Arm", which is pretty much the same thing as clicking the safety off on a loaded gun. My weapons were now live, and with the press of the pickle-button or the squeeze of a trigger I could rain hate on the enemies of mankind. I looked over at Colt 44…he was still in perfect formation. Checking my navigation, I realized we were about to cross over into Kosovo's airspace. I keyed the mic again to make sure Colt 44 was on the radio and to contact the NATO AWACS, "Colt 43 check." Colt 44 crisply responded, and I continued, "Magic, Colt 43 checking in as fragged." "As fragged" meant that everything was in accordance with the schedule, from the number of and type of aircraft in my formation, to what weapons we were carrying, and what mission we were flying. After a pause while the AWACS found me on his radar scope, the controller replied, "Magic copies, Colt 43. Picture clear," letting me know that there were no enemy fighters for me to worry about.

As we cruised over Kosovo to meet up with Bear 81, I looked down at the countryside below. The terrain was rugged to rolling, with some pretty high mountains, still capped in snow. It was actually very picturesque, and I had to remind myself that down there, people were killing other people simply because they were different from each other. I also reminded myself to keep weaving my jet gently from side to side so I could watch those mountain peaks; they were high enough that a bad guy on top of one with a man-portable SAM could shoot me down even though I was at 25,000 feet. As I navigated my formation to Bear 81's position, I settled into my combat cross-check. I spent most of my time looking outside, checking the sky around us for threats with an occasional pass through each of my sensors and the radar to detect things my eyes could not see. As we approached the target area Colt 41 and 42 contacted me on our private flight radio frequency. They were low on gas, so they began heading back to the tanker since we were close enough now to protect Bear 81.

"Bear 81 this is Colt 43, on station," I called over the strike frequency. "Bear copies," came his reply. "Colt 43, my next set of fighters will be on station in about 10 minutes. Stay above 25,000 and you'll be out of the way." I clicked the mic twice to acknowledge and laughed out loud in the cockpit. The pilot flying Bear 81 was also another old friend from a previous assignment! Bear 81 and I had learned to fly F-16s together back at Luke Air Force Base when we first graduated from pilot training. He was a great pilot, and to this day is hands-down the nicest guy I have ever met. Later in our careers we would serve together again when I was a squadron commander, and he was my second in command.

Still shaking my head at the coincidence of running into two different guys I flew with before on this mission, I maneuvered our flight into an orbit near the target area. The trick was to stay close enough for us to cover Bear 81 and the incoming fighters, but distant enough that we wouldn't be in their way. The target area itself was in the vicinity of a small town with a name that nobody could pronounce, so it was called "D-Town" for short. As I circled, I looked down at the scene below. There was a small hamlet a few miles outside of D-Town, and smoke was rising from several of the houses that had been set on fire. Since I was well clear of Bear 81, and the next flight of bomb-laden fighters hadn't shown up yet, I decided to drop down lower for a better look. I told Colt 44 to stay put, turned my transponder off so the AWACS would have a hard time seeing what I was up to, and I rolled my jet upside down and performed the second half of a loop (a maneuver called a "split-S," because it looks like the bottom-half of the letter "S"). The ROE prohibited us from going below 15,000 feet in most circumstances, but I didn't figure anyone would notice my momentary deviation. I left the throttle all the way up just short of afterburner as I pulled out the split-S and rocketed across the town at 1000 feet and 600 knots. My heart sank at what I saw flash beneath my jet.

The Yugoslavian forces had rolled a military convoy into the hamlet sometime earlier in the day and were burning the place down house by house. I could see the Yugoslavian troops milling around, and large groups of civilians herded into nearby fields. I knew what was going on from open press reports we had seen on TV. The Yugoslavian troops had rounded up everybody in the town at gunpoint, and then separated the civilians in three groups: military aged males, women, and children. According to the media accounts, the military age males would be executed, the children sold off into servitude, and the women sent to rape-camps. I didn't expect to ever see something like this myself. Sadly, the ROE was written in a way that prohibited me from doing anything to stop them. At that moment, I heard the fighter-bombers check in, so eased the stick back and zoomed my F-16 up to rejoin with Colt 44.

The fighter-bombers checked in with Bear 81. They were a flight of two F-15Es, callsign Shark 21. Shark was loaded down with GBU-12s, which are 500-pound laser-guided bombs, and they had plenty of gas to stay over the target area for as long as Bear 81 needed them. Bear 81 had located a Yugoslavian tank rolling along a tiny road through a wooded area and had been able to check off all of the things in the ROE required to declare the tank as a hostile target. What followed was one of the most unbelievable sequence of events I've ever witnessed in hundreds of hours in combat.

Operation Allied Force was a NATO war, and as such, the ROE had to be agreed upon by all of the countries involved. In addition to a generic set of standing ROE, each country had its own nation-specific ROE that could be more restrictive than the general ROE. Given the propensity of politicians to dictate military tactics, it should come as no surprise that the U.S. ROE had some really quirky things in it. One of them was the definition of what constituted killing a tank. Basically, we could only report a tank as destroyed if the turret was separated from the chassis. So, Bear 81 talked Shark 21 onto the tank, and cleared him to drop a single GBU-12. Shark 21's bomb guided right onto the target, and the shock wave from the blast rippled through the trees and blasted a ton of debris into the sky. For several minutes we continued to orbit, waiting for the smoke to clear. When it did, Bear 81 dropped down to 15,000 feet and made a pass over the tank. "Shark 21," Bear 81 called over the radio, "You hit it, and it's not moving anymore, but the turret is still on it. Let's have Shark 22 give it a try."

So, we went through the whole drill again. The smoke cleared and Bear 81 checked on the tank a second time. "Shark 22, you hit it and turned the whole tank around 180 degrees, but the turret's still in place. Shark 21, you get a second try." Shark 21's bomb was good again, but this time the tank was turned over on its side...but the turret was still in place! Shark 22 dropped a second GBU-12, and this time turned the tank back upright. "And Shark 22 you won't believe this, the tracks are off it, but the turret is still on there." "Okay," Bear 81 continued, "Shark 21 and 22, I'll keep alternating you both on the target, and I will buy a beer for whoever blasts that turret off first." Fighter pilots are a competitive bunch of guys, and a chance to win was like, well, blood in the water for Shark flight! Time and time again they took turns dropping on that tank. The tank bounced all over the place. Off the road, back on the road, and at one point it was even bounced upside down before being blown back upright by the next bomb. But no matter what, that damn turret wouldn't separate from the chassis! Bear 81 was having a blast (literally and figuratively) plinking that tank with 500-pound bomb after 500-pound bomb. Both Sharks were getting angrier and angrier that they couldn't win the beer. I was literally laughing until I cried, imagining that tank, stripped of everything but the turret by the crushing explosions, the crew inside smashed into jelly, and still, we couldn't declare the tank dead in accordance with the ROE! You just can't make this stuff up!

It was the only target Bear 81 could find, so Shark kept dropping and the tank kept bouncing. Eventually Bear 81 ran out of gas and had to head home. "Shark 21 flight," Bear 81 called, "I can give you 100% of bombs on target for your strikes, but for BDA (Battle Damage Assessment, which is what you achieved with your ordnance) I can only give you one tank damaged. Thanks for the good work, and I will buy you both a beer anyway!" Shark sounded like he was sulking as he acknowledged and turned his flight for home as well. I checked our fuel, and it was time for us to go back to the tanker and get a top-off before our next mission. F-16CJs were in high demand, so it was common for us to have to provide cover to several missions while flying a single sortie.

I led Colt 44 back to the tanker track back over Albania. As I drew closer, I could see that there was another flight of F-16s already getting gas. As I got closer, I could see from the tail-flashes that the jets were other F-16CJs from Spangdahlem Air Base. I decided to hang back a few thousand feet until they were clear. As the last Spangdahlem jet cleared the tanker the flight lead, Kimber 51, came up on the radio. For the third time that day I recognized the voice of a pilot I had flown with in the past! I asked Kimber 51 for his inter-flight radio frequency and gave him a call. My old friend and I chatted for a few minutes. He, and I, and Bear 81 had all learned to fly F-16's together in the same class at Luke AFB. Kimber 51 had been providing SEAD near Bear 81 earlier and recognized his voice on the radio too, and we had a good laugh about all of us flying a combat mission together. "What are the odds?" I asked Kimber 51. "I don't know," he replied, "but it's like old-home week up here!"

Kimber 51 left the tanker, and I turned my attention to getting gas. Before I knew it, we were fencing-in again for our second mission that day in the KEZ. This time around

I was supporting another FAC, an A-10, callsign Nail 21. He was working an area about half-way between D-Town and the capital of Kosovo, the city of Pristina. Unlike Bear 81, Nail 21 had found several targets, and had several flights of fighters (callsigns Rolls, Dragon and Hatchet) stacked up, patiently waiting their respective turns to drop their ordnance. As Nail worked the fighters over the target, Colt 44 and I sauntered about nearby in case we were needed. I positioned our orbit so that half of the time we were watching the area under the FAC and fighters for SAMs and AAA, and the other half of the time we were looking at Pristina airport. In the past, the Yugoslavians kept fighters at Pristina, so we wanted to be on the lookout for any MiGs that might want to come up and play. After the entertainment of watching Bear 81 and Shark flight make that tank dance, the attacks Nail 21 controlled were downright anticlimactic.

As the time approached for us to go home, I received a call on the radio. "Colt 43, Nail 21 here. I just found some guys loading missiles or rockets of some kind onto a launcher. Can you guys stick around for an extra 15 minutes?" This really perked me up. If there was a launcher and it was being loaded with SAMs it could mean things were about to heat up. I told Nail 21 that we were happy to stay, and I asked him for the coordinates of the launcher. After copying the coordinates, I loaded them into my navigation and targeting systems while I listened to Nail 21 coordinate an attack using Hatchet flight, a two-ship of Navy F-18s. As I repositioned my wingman and myself to cover Nail and Hatchet's attack, I started getting indications on my equipment that several Yugoslavian early warning radars were active somewhere to the north of us.

Early warning radars are used to detect the general location, altitude, speed, and heading of airplanes from long range. This information is then passed to command-and-control nodes who process the information, and frequently pass it to the SAM and AAA batteries who use that data to queue their weapons. This does two things for the enemy. First, by minimizing how long they kept the SAM and AAA radars radiating, the enemy decreased our likelihood of finding them and killing them before they could shoot us. And second, by remaining silent until they are ready to shoot, they maintained the element of surprise. In turn they could wait until we were in the heart of their missile and AAA engagement envelopes before attacking. The same early warning radar data could be used to vector MiGs on us as well.

All of these things passed through my mind as I set up to support the strike that was unfolding. I maneuvered my flight away to the north while the FAC and the strikers were outside of the range of any SAMs or AAA in the target area. This gave me a good chance to watch the early warning radars, as well as look for any MiGs that might try to sneak up while the attack was in progress. When I heard Nail and Hatchet call that they were beginning their attack, I wheeled my wingman and I around to follow. My timing was perfect, and as I rolled my jet out of the turn, I was saddled up 15 miles behind Nail. This provided me with plenty of time to look for SAMs and AAA in the target area or underneath their approach.

As we pressed into the target area, I smoothly scanned the sky around us, the ground beneath us, and each of my sensors in turn. Although I saw no indications of a threat to Nail and Hatchet, I kept an AGM-88 HARM ready, with my thumb poised over the pickle-button just in case a SAM or AAA radar popped up. Over the radio I listened as Nail 21 cleared the F-18s "hot," meaning they were authorized to drop their bombs. Hatchet called off target, and I checked my jet a little to one side so that I could look around the nose and see the target area more clearly; about 30 seconds later I saw Hatchet's bombs detonate, the shockwave rippling across the ground. F-18s are fast jets, and Hatchet flight was clear of the target area quickly. Nail, being an A-10, took a little longer to leave the target area, and being at the end of his mission called it a day. Colt 44 and I stayed in the area until we were the last ones in the NFL. After such a promising start, this strike was anti-climactic. That's how a lot of combat missions go, thankfully, I suppose. But you can never let your guard down. You can never assume the enemy is asleep, or there is no threat; that's how you get yourself or others killed. From the time you fence in, to the time you fence out, you have to stay focused, disciplined, and switched-on.

My wingman and I fenced out of bad-guy land and headed back to the tanker to get enough gas to fly back to Aviano. As we flew north, up the eastern coast of the Adriatic Sea, I looked down at the rocky beaches, and inlets, and small islands dotting the shoreline. It was really beautiful. It would probably be a great place to vacation if everyone wasn't trying to kill each other over their ethnic and religious differences. An hour later I climbed out of my jet after a nearly seven-hour sortie. The sun was starting to drop low in the sky, and after thanking the crew chief for his hard work, I slung my helmet bag over my shoulder and waited for the crew van to pick me up and take me back to our operations building. The adrenaline of the day was starting to wear off and I was tired, and very, very hungry. My thoughts turned to what I would have for dinner?

Chapter 5: Distinguished Flying Cross

A few days later I was on the schedule to fly a particularly interesting mission. About a week prior, the Serbs managed to shoot down an F-117 Nighthawk with an SA-3, like the one I dodged at Obrava. At the time, nobody knew how they had managed to detect and target a stealth fighter. Maybe it was just a lucky shot? Or perhaps the F-117 had a panel fall off, stripping away some of its stealth coatings and making it visible to radar? Or maybe the Russians or the Chinese had found a way to detect the F-117 and were helping the Yugoslavians? And if they could see an F-117, did that mean they could see a B-2 Stealth Bomber as well? Nobody knew for sure, but the U.S. air war planners weren't going to take any chances. After much deliberation it was decided that the F-117s and B-2s could resume their role in the bombing campaign, but only with F-16CJs as an escort. The thinking was that not only could we detect and shoot the SAM and AAA radars when they came up during the air strike, but because our jets were decidedly un-stealthy, the Yugoslavians would target us instead of shooting at hard-to-find stealth airplanes.

I always enjoyed night flying during Operation Allied Force, for the simple reason that it was much easier to see the SAMs and AAA when they were shooting at you. Rockets motors, tracer rounds, and air-bursting munitions really stand out at night. And considering it was our job as Wild Weasels to find these weapons and destroy them, being able to see them was a real plus. Since the whole mission would be at night, we didn't arrive at the squadron until after dinner.

Our callsign that night was Cobra, and I was flying as Cobra 43...the number three man in the four-ship. We went over to the F-117 squadron for the briefing, and after a weather and intelligence update, the mission itself was covered in great detail. We (the F-16CJs) would be taking off out of Aviano AB and flying east across Croatia and into Hungary. Once there we would refuel with a KC-135R just south of Budapest, and then proceeding south, across downtown Belgrade, and setting up our CAPs on the south edge of the city. From that position we would be well-positioned to cover the F-117 strikes. After the F-117s were clear of the target area our flight would transit back across Belgrade to the tanker track in Hungary and refuel again. Once we received our gas the plan was to fly south yet again to Belgrade and cover a B-2 strike. Once the bombers were safely away, we would gas up once again in Hungary before flying back across Croatia to Italy. Because of the highly sensitive nature of F-117 and B-2 operations, these kind of strike packages were called "U.S. Only" missions, and we would only be working with U.S. AWACS and tankers.

After the mass briefing, we held our own four-ship flight briefing, covering the specifics of what we would do to support the F-117s and B-2s. It was a complex problem to solve. The skies over Belgrade were the most heavily defended airspace the U.S. military had faced, with scores of SAM batteries and hundreds of AAA guns. Belgrade, being the capital city, was also surrounded by several MiG bases, which meant that we would need to be ready for MiGs to launch into the middle of things with little or no warning. Because of the MiG threat, I was especially happy to have a USAF AWACS covering us. I really didn't trust the NATO AWACS at all, after some of the things I had seen and heard them do over the years.

Basically, after looking at the routes the stealth guys would fly and the threat laydown, we picked a spot over south Belgrade that would put us in the best location to keep an eye on everybody. We would split into two flights of two F-16CJs, with our orbits about 15 miles apart. Geographically separating our orbits was important to make sure we didn't run into each other in the dark. Believe it or not, F-16CJs didn't, at that time, have night vision goggles (NVGs). Once we fenced-in we would turn out off our exterior lights to make our airplanes (hopefully) invisible to people on the ground looking to shoot at us. The side effect of this was that we couldn't see each other, either! To prevent midair collisions, we always deconflicted our flight paths geographically, as well as by flying at different altitudes. Once on station we would coordinate our orbits so that at any given moment, two of us would be looking at the target area, and the other two would be looking behind us. As Cobra 43, I planned to take Cobra 44 to an orbit east of where Cobra 41 and 42 would be. We would then be able to coordinate between the four of us to shoot any radars that threatened the stealth guys, and also be in a good position to keep our air-to-air radars on the many different MiG bases surrounding us.

We would go to the same location for the B-2 strike when that occurred.

Staring at the map, I mused that Cobra 44 and I could end up relatively close to the border with Romania. The prevailing winds at our flight altitude would be blowing west to east at around 100 knots that night. This was important to take note of, as making for Romanian airspace would be a good plan if a jet were to be hit by AAA, or a SAM, and couldn't stay airborne for long. Romania was a neutral country in this conflict, and while they might not return us to U.S. custody right away, staying a guest in Romania was a much better plan than becoming a prisoner of war in Yugoslavia. We had a good laugh when Cobra 11 pointed out that the nearest point of crossing would be the Transylvanian Mountains…home to Count Dracula's Castle! Too cool! After our briefing we went to the Life Support building to get our flight gear. And here is where I made the first big mistake of the night…one that almost cost me my life!

As I mentioned earlier, F-16CJs did not have any special night fighting capability. We didn't have NVGs, our cockpits did not have special lighting to keep your night vision sharp, and we didn't have targeting or LANTIRN pods to help us turn night into day either. Basically, we flew and fought at night just like the guys did in WWII. Even though we had great air-to-air radars, we couldn't use them to their fullest effect to maintain formation. We needed the radars to search for MiGs, so we really couldn't spend too much time using them to watch each other. Besides, locking someone with the radar when we were all caring live weapons would be a bad idea. It would be kind of like walking around in the dark with a loaded gun pointed at the head of the person in front of you. Only bad things could happen. Years later, the F-16CJs were equipped with NVGs, datalinks, and other cool equipment to make night flying easier. The young pilots would shake their heads in amazement that in the "old days" we could execute formation tactics at night without seeing each other. But that was all we had, and we did just fine!

So, given the dearth of night-specific equipment, you can imagine my curiosity when the Life Support Technician asked if I wanted a "finger light." I said yes, not really knowing what it was, and I was even more puzzled when the technician handed me a 35mm film container. As he walked away, I cautiously opened it and shook the contents into my hand. The finger light was a little rectangular box with a small square LED housing on one end, with a 3–4-inch strip of Velcro attached to it. I held it up in turned it around a few times, perplexed as to what this was and why was it called a "finger light?" Cobra 44, my young First Lieutenant wingman, chimed in with an explanation as he stuffed his finger light into his helmet bag. "The guys who fly with NVGs use these." He took it from my hand and wrapped the Velcro strip around his pointer finger with the LED housing perched on top of his knuckle pointing away from him. With his other hand he reached over and depressed the LED housing back into the rectangular box, and with a small "click" the LED lit up, projecting a narrow green beam of light. My young Lieutenant wingman waved his finger around adding, "Since the green light doesn't hurt your night eyesight, you can use this. You just turn off all the cockpit lights, and then you point the light at whatever instrument you want to see." "Cool!" I replied, taking the finger light, and stuffing it into my pocket. "Thanks for showing me how to use it!" I didn't give it a second thought, but I should have known better than to trust a Lieutenant for advice when flying in combat at night.

Soon we stepped out to the jets, and after an uneventful preflight, engine start, and taxi, we were quickly airborne and winging our way across the Adriatic Sea to Croatia and beyond. I have always enjoyed night flying, regardless of whether I was on a combat mission or a training sortie. There is something uniquely comfortable about being settled into the little world of single-seat fighter at night…it's like being in a cocoon. I began sorting out my cockpit in anticipation of the night's work when we were about 15 minutes away from the tanker. I drank an entire bottle of water, took a leak, and then quickly reviewed my mission materials to refresh the different codewords and callsigns for the night. It was then that I remembered the finger light.

I dug around in the leg pocket of my G-suit and found it. After strapping it to my finger I turned it on and pointed it around the cockpit. Following the Lieutenant's instructions, I turned off all of the cockpit and instrument lights. The cockpit was absolutely pitch black, and I couldn't see anything unless I pointed the finger-light right at it. In a few minutes I felt like I had the hang of using it. With it strapped to my left forefinger, I could keep my hand on the throttle, and point the light at each flight instrument around the cockpit as needed. With no white lights in the cockpit, my eyes adjusted quite well to the darkness. "This is awesome!" I thought to myself, "How come we never used these before?" By this time, we were approaching the tanker, and I turned my focus on the tasks at hand, in a completely blacked-out cockpit.

Everything was going pretty smoothly. We got our gas, checked in with the U.S. AWACS, callsign Darkstar, and fenced in. As we crossed into Yugoslavia, I looked out at the city of Belgrade. It seemed unreal to me that we were bombing a European capital...something that hadn't happened since WWII. Equally unreal was that the lights were still on in Belgrade. You would think that one of the first things we would have bombed was the city's electrical grid. Turning out the lights in Belgrade would have definitely made things harder on the Yugoslavian regime, but for some reason we didn't do that. I guess NATO was too concerned about the effects of no electricity on the civilian population. Even more amazing, was the fact that the Yugoslavians left the lights on when they knew we were coming to bomb them. With the city lit up like, well, a major city at night, it made it really easy to navigate around and find our targets! As we approached the city, I began to scan my sensors and the sky around me in earnest. It was showtime.

We crossed over Belgrade without any real excitement. The Yugoslavians were watching us with several early warning radars, but they hadn't turned on any SAM acquisition or guidance radars. We didn't even see any AAA. "So far so good," I thought. At the appropriate time and place I turned me and my wingman to the east and set up in our orbit. I was really in the zone and switched-on, constantly interpreting each of my sensors, the radar, and scanning the sky and ground for any activity. I was flying at 25,000 feet. My wingman was saddled-up behind me about three miles back, offset to one side by about 30 degrees, and flying 2,000 feet higher than me. In theory, this would help to keep us from running into each other as we flew around in the dark with no external lights on. By design, we had arrived in our orbits only minutes before the F-117s would arrive. This would give us just enough time to study any radars that came on, compare their locations and lethal ranges to the flight path of the F-117s, and shoot the ones that were most problematic.

In combat, luck (both good and bad) certainly plays a big role. Just moment before the F-117s were supposed to begin dropping on their targets, the Yugoslavian air defenses opened up and came alive. In a stroke of bad luck, Cobra 41 and 42's western orbit was directly above a mobile and highly lethal SA-6 SAM battery. The bright orange glow of the first SAM launch immediately caught my attention. Separated by a few seconds, two missiles streaked upwards. Cobra 41 made a quick call that they were defensive. I immediately rolled my jet up on its wing and made a high-G turn to engage the SA-6. I intentionally didn't use my afterburner because I didn't want the giant plume of flame it would generate to give away my position. The tradeoff was that my heavily laden F-16, flying at 25,000 feet, lost a lot of airspeed in the turn. Rolling out, pointed directly at the SA-6 site, I heard Cobra 42 call that the missiles had overshot. As I was about to fire a HARM at the SA-6 my sensors detected an SA-3 in Belgrade, and from its position I knew it was a threat to the F-117s. I only had a split second to decide...should I go after the SA-6 shooting at Cobra 41 and 42, or should I turn and engage the SA-3 that might shoot at the F-117s? F-16s are far more maneuverable than F-117s, plus it was our job to protect the guys dropping bombs, so I decided to go after the SA-3.

Once again, I rolled my jet up and, without using the afterburner, made another high-G turn to engage the SA-3. As I rolled out, I saw two more SA-6 missiles launch at Cobra 41 and 42. I could see the afterburner plumes from both F-16s as they began to dodge the missiles. At that moment my wingman called that he had detected another SA-3 that was a threat to the F-117s, and that he was attacking it. "Damn," I thought to myself, "this is getting serious!" I knew I had to help Cobra 41 and 42 as soon as I dealt with the SA-3 I was targeting. Focusing on my HARM Targeting System, I locked an AGM-88 HARM onto the SA-3. I glanced at my Master Arm Switch to make sure my weapons were live, but to my dismay I couldn't see the switch…I had turned off all of my cockpit lights and was depending on the finger light! Double checking that the Master Arm Switch was on was just that, a double-check. I was confident that I had turned it on earlier, so without wasting another second, I moved my thumb to the "pickle" button on my stick and fired the HARM.

Of all the weapons I have fired or dropped from an F-16, launching a HARM is the most magnificent. The HARM is a big missile. When the rocket motor fires, the thrust is so great, that the F-16 actually yaws as the missile breaks free from its launcher rail. The rocket plume is the purest bright-white light, and as the missile streaks away it looks like a meteorite. If you happen to be looking in the cockpit when it launches, you will see your engine RPM roll back a few percent, as the engine ingests the hot gasses from the HARM's rocket motor. And it is fast…the "high speed" in High-Speed Anti-Radiation Missile is no exaggeration!

But on this occasion, I didn't have time to admire my work and watch the HARM do its thing. I needed to help Cobra 41 and 42.

Once again, I rolled my jet up into a high-G turn to re-engage the SA-6 beneath Cobra 41 and 42. In the middle of my turn I saw two more SA-6 missiles rise up towards them, and Cobra 42 called that he was performing a "last ditch" maneuver to avoid them. Things were getting really bad for Cobra 41 and 42, and I was sure that they would only be able to dodge a couple of more missiles before one of them got hit. But before I completed my turn, I saw the flash of a missile launch on the ground directly under *my* jet. Things had instantly taken a turn for the worse, as I realized that I too was directly over an SA-6 site. In an instant, I knew that the Yugoslavian SAM operators must have seen me launch my HARM. The bright rocket motor would have given the SA-6 crew a visual location to begin cueing their radar and launching missiles.

A cold reality hit me at that moment: after zig-zagging my jet back and forth, at high-G without using the afterburner, my airspeed was really low. Low airspeed, especially at 25,000 feet and heavily loaded with weapons and external fuel tanks, meant that I could no longer turn the jet very fast. With no ability to turn the jet quickly, I was a sitting duck for the SA-6. With what little airspeed I had left, I rolled my F-16 away from where Cobra 41 and 42 were dodging missiles, so that I wouldn't make matters worse for everyone by flying into the middle of their drama. As soon as I was pointed away from Cobra 41 and 42, I pulled the nose of my jet straight down to get some airspeed, and simultaneously create a new geometry problem for the missile guiding on me to solve. My eyes were locked onto the SA-6 missile as it rocketed toward me, and I activated my electronic counter measures and began dispensing chaff in an effort to decoy the missile away. Knowing I would only get one chance to dodge this missile, slammed the throttle into full afterburner and reached for the Emergency Jettison button to drop all of my external stores. With no missiles or gas tanks hanging on the jet, I would be able to accelerate more quickly and make a harder turn to dodge the missile at the last second. At that very moment my blood froze…I had turned off all the cockpit lights, and now I couldn't see the Emergency Jettison button without taking my eyes off the SAM and directing the narrow beam of green light from the finger light! I realized I couldn't see my altimeter or my airspeed either. At least I could tell which way the ground was…because the AAA tracers were shooting up! "I deserve this for listening to a Lieutenant," I thought to myself.

The missile was getting very close, and for some reason I remembered that the Transylvania Mountains were nearby. I wondered if I might get a chance to parachute into Dracula's Castle. About this time, I noticed that there was a lot of AAA fire coming up at my jet. From the number and color of the tracers I knew it was 37mm and 57mm fire, which meant I was getting much lower to the ground. That was no surprise as my F-16 was hurtling straight down in full afterburner. What did surprise me was how accurate the AAA fire was. I thought for sure the AAA rounds were going to get me before the missile did. The tracers were going by my jet on all sides, and one string went by so close that I could hear and feel the supersonic shock waves of the shells passing inches from my canopy! In flash I was below the AAA and the SAM was on me. I pulled the stick back hard, trying to make the missile miss by every little bit I could. Later, when I was back at Aviano and watched my video tapes from the mission, I discovered that pointing the jet straight down in full afterburner had saved my life. When I began to evade the missile, I was going almost 450 knots. So, when I pulled the stick back as hard as I could, my F-16 gave me a solid 9-G turn. At the last instant, the

SA-6 missile passed behind and above my jet and exploded, missing by about 50 feet or so. It was close enough that the SA-6's proximity fuse sensed my jet, and I felt the shockwave from the detonation. It felt kind of like flying through someone's jet-wash.

And suddenly everything seemed calm for a moment. There I was, flying along straight and level in full afterburner. I looked around my pitch-black cockpit and was amazed that I didn't see any warning or caution lights illuminated, indicating battle damage. Somehow, I didn't get hit! "Holy shit," I yelled! But I knew I didn't have time to relax. Quickly assessing my situation, I had some hard choices to make. After the 9-G pull to avoid the missile, I was flying at 14,000 feet. Being that low put me in the heart of the engagement envelope for the SA-6, as well as the AAA. It was only a matter of time before they figured out where I was, especially with the plume of my afterburner blazing brightly in the night sky. If it had been daytime, I would have done what I did a few days earlier and dropped down as low to the ground as I could and run away at high speed. But without NVGs it would have been suicide to try that in the dark. So, my options where to keep speeding up and run outside of the threat area or try and get some altitude back and get above the AAA.

Honestly, at this point I was far more afraid of the AAA than the SA-6, because whoever was shooting those guns at me had pretty good aim! So, I decided to go for altitude first, get above the range of the AAA, and then worry about the SA-6.

As I mentioned before, the GE-129 engine in the F-16CJ Block 50 is wondrously powerful. Leaving the afterburner lit, I pulled the nose up until I was climbing at 200 knots. My jet climbed like it was on an elevator, all the while I was twisting around in my seat looking over my shoulder for more SAMs and AAA. I saw more SAMs being shot at Cobra 41 and 42, and another SAM being fired from nearby, but it was clear that it wasn't aimed at me.

Still looking over my shoulder, I was suddenly blinded by the afterburner plume of a jet passing me close by, headed downward. It was my wingman, Cobra 44, dodging a SAM that was fired at his jet that I had not seen! I quickly glanced at my Head's Up Display and saw that I was passing 20,000 feet, and my jet was slowing down, so I pushed the stick forward and bunted back to level flight. As I accelerated out to a tactically sound airspeed, I reassessed the situation. I heard Cobra 41 and 42 call that they were no longer defensive. I guess the SA-6 battery had run out of missiles. Cobra 41 and 42 were, however, in an emergency fuel state, so they immediately left their orbit. They couldn't go back to the tanker because they would have to fly through the middle of the F-117s dropping their bombs. Instead, they knew they had just enough gas to make it back to Aviano, and that's what they did. For them the mission was over. My wingman, having successfully survived his encounter with an SA-6, was climbing back into formation with me. And for some reason, the AAA and SA-6s beneath us had gone quiet. "Okay," I said to myself, "time for a new plan."

The first thing I did was turn up the cockpit lights. Then I pulled out my map and checked my watch. Noting the time, I compared it with the time over target for the different F-117s. With Cobra 41 and 42 off station, I needed to move to a new position where I could cover the remaining F-117 strikes with just me and my wingman. I decided to fly back across the top of Belgrade, climbing above the F-117s, and set up shop on the north side of the city. That would put me in a position to engage any SAMs threatening the remaining F-117s, watch the most likely MiG bases with my radar, and put me closer to my tanker up in Hungary. Dodging those SAMs and the AAA had eaten a lot of gas for both me and Cobra 44, but I calculated we could finish the mission and still make the tanker with a little gas to spare. With Cobra 44 back in formation, I turned us northwest.

As we flew across Belgrade, I didn't see any more SAMs come up, but there was a huge volume of AAA. Fortunately, we were now back above 25,000 feet, so most of it wasn't a threat to us, but it was still eye-watering to see dozens and dozens of ropes of tracer fire hosing back and forth over the city. The heavy caliber 100mm AAA that the Yugoslavians had did not use tracer rounds, but I didn't see any of their characteristic white airbursts, so I assumed they weren't shooting at the moment. The Danube River looked like a black ribbon weaving through the city lights, and that made it easy for me to navigate to our new orbit. I also took a moment to tell Darkstar that I was moving us to a new location on the north side of the city.

When we reached our new orbit, I turned us back toward the center of Belgrade and resumed our coverage of the F-117s. Rolling out and looking at the city before me, I was shocked to see the volume of AAA increase even more. It looked like the finale of a July 4th fireworks display, with the sky brightened by exploding shells and tracers. At that moment, the AWACS made a call that peaked my attention. "Cobra 43, Darkstar, hostile groups maneuvering over Batchinika Airfield." "Oh BABY!", I thought, "Tonight's the night!" Dropping bombs and shooting HARMs is fun, but air-to-air combat is the sport of kings. My attention immediately dropped to my radar scope. I moved the cursor on my radar over Batchinika Airfield. The whole radar scope was cluttered with contacts at that location, moving in all directions. I picked one and locked it up with my fire control radar. Immediately, I realized that something wasn't right. The contact I had locked was dropping straight to the ground at a slow speed, and my radar quickly broke lock. I tried again with another contact, and it appeared to go up for a second before heading down and breaking my radar's lock once again. Batchinika was only about 10 miles away, so I looked over the nose of my jet to see what my eyeballs could tell me. I could see Batchinika quite clearly, and immediately I understood what was happening. There was so much AAA fire coming out of Batchinika, and detonating over the airfield, that the volume of shrapnel was creating a cloud of metal that our radars could see! Darkstar figured that out on their own about the same time and let me know they no longer saw any enemy fighters. I won't lie, I was disappointed. I wanted to kill some MiGs very badly.

I didn't have time to dwell on my disappointment for long. At that very moment my RWR lit up indicating that an SA-6 was locking onto my jet. At the same time Cobra 44 also called that he was getting engaged. I rolled my jet up on its side to look underneath just in time to see SAMs launch off the ground from two different, but close together, locations. In another stroke of bad luck, the random place I had chosen for our new orbit was next to another SA-6 battery! But this time both my wingman and I were in a much better position to defeat the SAMs, with enough altitude to stay above most of the AAA as we maneuvered, and plenty of airspeed to keep maneuvering. Watching the two missiles tracking toward us, I pushed the throttle into full afterburner, deployed chaff, activated my electronic countermeasures, and began turning my jet to create intercept problems for the missiles solve. Cobra 44 was doing the same thing in his jet. I rolled inverted and began tracking my nose down below the horizon and then rolled back upright, all the while never taking my eyes off the SAMs. My maneuver confused the missiles for a second, and I watched the SAMs as they tracked away from my jet momentarily. Soon, however, they began to turn back toward me, so I pulled the nose up above the horizon to create another change of geometry for the missiles to solve. This time the missiles took even longer to correct back toward me, which was just what I wanted. I was getting the missiles to maneuver out of synch with my airplane. In just a few seconds it became obvious that I would defeat both SAMs, and it was a great relief when they both passed harmlessly behind and above my F-16. Cobra 44 defeated his SAM at about the same time. No more SAMs came up and the AAA beneath us seemed to slack off as well, but now my wingman and I had a new problem…we were out of gas.

Using the afterburner in an F-16 uses lots of gas. When we moved to our new orbit over the north side of Belgrade, we didn't have a lot to spare, and using the afterburner to dodge the most recent SAMs now meant that we didn't have enough gas to make it to a tanker. I thought about things for a second. The last of the F-117s was coming off target around this time, but with so much AAA over Belgrade I was quite sure they would be hidden in the radar clutter created by all of that metal in the sky beneath them. Over the radio I confirmed Cobra 44's gas totals, and they were identical to mine. I did some quick math and decided that with a little luck we should be able to just make it across the border into Hungary before our engines quite due to fuel starvation. I thought about jettisoning our external stores to get a few extra miles, but we were only a few minutes flight time from the border, so I decided the risk of the Yugoslavians recovering our gear wasn't worth it. As we turned north for the Hungarian border, I called Cobra 44 on the radio. "Look at the bright side. If we make it to Hungary, we can at least walk home!" Cobra 44 came back, his voice cool and collected, "And we won't be guests of Count Dracula tonight."

So, with that moment of levity, I turned my focus on how to maximize our chances of getting to the tanker. The first thing I did was call Darkstar and let him know our fuel situation, and that in all likelihood we would be ejecting before we made it to the tanker. After Darkstar let me know that he would alert the Combat Search and Rescue (CSAR) forces to get ready, I asked him to see if the tanker could fly south and meet us right at the border. I figured it was worth a try. We were certainly due for some good luck after everything that happened over the last 20 minutes or so. I slowed down to an airspeed that would give us the maximum range with the gas we had remaining. Then I turned my attention to stowing everything in the cockpit so that when I ejected, I wouldn't get killed by an errant water bottle or something else stupid like that. Next, I tightened the straps to my parachute harness, as well as the chinstrap on my helmet and oxygen mask. After that, I really didn't have anything to do but sit there and see how things played out. Honestly, sitting there waiting to eject was worse than getting shot at.

As we approached the border, I began to feel a little more optimistic. We were doing a tiny bit better on the gas situation than I expected, and I could see the tanker coming toward us on radar. We fenced us out and I switched us over to the tanker's frequency. They already knew our situation and would do everything they could to get to us in time. Normally the tanker just flies around in a predictable orbit, and the fighters maneuver to rejoin on him. This time, we just flew straight and level and the tanker timed his turn to roll out in front of us, and they did it perfectly, arriving wings-level in front of me only a mile away. Maybe our luck was getting better!

As we closed on the tanker, I checked Cobra 44's gas again; we were both within gauge-tolerances of empty. I then told my wingman I wanted him to get gas first...but only enough to buy him some time so I could get some too. I let Cobra 44 slide into the lead and followed him just off his left wing. He would only have one shot to get hooked up to take some gas, and I was worried the pressure of the moment might give him some trouble flying steady enough behind the KC-135R. But it turned out he was ice-cold, and hooked up to the tanker so smoothly you would have thought it was a training mission and not a combat emergency refueling. The KC-135 pumps about 2,500 pounds of fuel into an F-16 per minute, and with our external fuel tanks a full fuel load would be around 12,200 pounds. I glanced at my fuel gauge…it was showing between 300 and 400 pounds. True to the plan, my wingman only stayed on the boom long enough to take about 500 pounds. As he backed away from the tanker, he slid over to the right, and I simultaneously moved to the boom. I was expecting the engine to flame out at any second, but I knew I had to take my time and do this right the first time. I wouldn't get a second try. I inched forward, with only 1 or 2 knots of overtake on the boom. The refueling receptacle on the F-16 in on the top of the jet and just behind the cockpit. I realized I was holding my breath as the boom passed by my canopy. I let out a big sigh of relief as I felt the "thunk" of the boom locking into my refueling receptacle. I was going to pull this off! I watched my fuel gauge start to increase and decided to take 1,000 pounds of fuel. As soon as I had that, I disconnected from the tanker and moved over so Cobra 44 could hop back on. This time he took a few thousand pounds of gas, and it was my turn again. I topped off my tanks, and then my wingman did the same…we were good to go!

Literally. Because it was time to head back to Belgrade and cover the B-2 strike that was inbound. Cobra 44 and I climbed away from the tanker and turned back south, fencing-in and contacting Darkstar to let him know that we were on our way back, and to get the status and location of Cobra 41 and 42. It turned out that Cobra 41 and 42 were currently enroute back to Aviano AB, so we would be on our own for this next part of the mission. Once again, I scrutinized the B-2 targets and timing, and adjusted where we would set up our orbit, given that there were only two of us now. I picked a spot smack-dab over the center of Belgrade. I figured that would be like giving the middle-finger to the Yugoslavian air defenses, and they would focus on killing us instead of searching for stealthy airplanes. After all, it worked earlier. As we approached the border, I took my helmet off for a moment, enjoying the cold air on my sweaty scalp. I looked around for a moment and decided to guzzle down a whole bottle of water. Feeling refreshed, I put my helmet back on and got ready for round two of this fight.

I was really jazzed-up as the city lights of Belgrade slid under the nose of my jet. I anticipated that the next 20 minutes or so would be pretty wild. The Yugoslavian air defenses had put up a huge fight earlier, and now we were coming back for more, after kicking the hornet's nest.

But nothing happened. Nothing. We cruised around over Belgrade for 20 minutes and not one shot was fired by us, the Yugoslavians or the B-2s. I stared at the B-2 targets, waiting for them to blow up, but no bombs went off anywhere. For some reason the B-2s aborted and Darkstar never received the word to pass along to us. So, Cobra 44 and I flew around over Belgrade like we owned it, bewildered that nobody was shooting at us. We didn't even see one string of AAA in the distance, or a tickle of a SAM radar anywhere. Nada. Nothing. For a little while I hoped the lack of AAA and SAMs meant that the Yugoslavians were sending up MiGs to fight us, figuring they didn't want to accidentally shoot down their own fighters. But that was wishful thinking on my part. It was a real letdown after all of the excitement during the first half of the mission. When our time was up, we went back to the tanker and then home to Aviano.

After I parked my F-16 and shutdown the engine, I post-flighted the jet. The crew chief and I gave it a good look over, because I was sure there had to be some holes in the airplane after all of the SAMs and AAA. But the jet was fine, with nary a scratch on it. While I was waiting for the crew van to pick me up and take me back to the squadron, I stood there with my helmet bag slung over my shoulder and looked at the F-16 I had just flown. It was perfectly silhouetted by the stadium lights surrounding the parking ramp, almost glowing in the light mist that was beginning to form. It reminded me of a dragon at rest. I was thinking about how much I loved being a fighter pilot, when Cobra 44 walked over from where his jet was parked, smiling ear to ear. "That was insane!" he said. I looked him in the eye for a second, and nodded, before reaching into my G-suit pocket and pulling out the finger light. I smiled and tossed it to him, "You can keep that, I think I'm better off without it!"

A few months later I was awarded the Distinguished Flying Cross for that mission.

Chapter 6: The Suicide Mission

Late one afternoon after I had flown a mission, I was hanging around the squadron and taking care of some paperwork until the chow-hall was open for dinner. Being a Flight Commander is a great job, but there was also a lot of drudgery involved in managing a dozen pilots. In the middle of reviewing the flight schedules for my pilots, our squadron Operations Officer plopped into the chair next to me. The Operations Officer, or "Ops O" is the second in command of the squadron. He made small talk for a bit, but I could tell he had something on his mind. "So, what can I do for you, Boss?" I asked. "Well, we have a pretty high-profile mission coming up tomorrow night. I want you to go over to the MPC (Mission Planning Cell) tonight and plan it, and then lead it. Put me on the schedule as your number 3. I would like to lead it myself, but I can't do the MPC tonight, and whoever leads this needs to be involved with the planning. Make sure we have solid wingmen too because we need an "A-Team" for this one." We chatted some more about how things were going in the squadron, and then he headed off to a meeting. I sat back and thought about what this "high-profile" mission might be. It was out of the ordinary for someone to go over to the MPC and plan their own mission. It was also very unusual for the Ops O to direct that two instructor pilots (IPs) would be in the same flight. We didn't have that many IPs, so we went out of our way to spread their experience around on the flight schedule. It must be a really special mission indeed! I took a look at the flight schedule for the next evening and re-arranged the names so that the Ops O and I were flying together. I selected a really experienced guy to fly as the Ops O's wingman, and then I pondered who I would have fly on my wing. I chose one of the youngest pilots from my Flight. He was the newest pilot in the squadron, and looked like he was about 12 years old, but he was really sharp and already had a reputation as a solid fighter pilot. I always felt a lot of responsibility for my wingmen, especially the young ones, and whenever I had the chance to schedule them to fly with me, I made sure it happened. This mission sounded like it could be especially challenging, but I figured he could handle it. Our "A-Team" was picked!

After dinner I headed over to the MPC. After the Air Tasking Order (ATO) dropped from Combined Air Operations Center in Vincenza, Italy, I learned more about this special mission. I was going to lead our four-ship of F-16CJs, callsign Cutlass 31, to support a series of strikes in and near Belgrade, and the city of Novi Sad. Novi Sad is across the Danube River and just to the northwest of Belgrade. The two cities are both huge and basically merge into one another. Belgrade and Novi Sad both had robust air defenses, with scores of overlapping SAM sites, as well as thousands of AAA guns. There were also several air bases around the two cities that hosted MiGs. It was the most heavily defended airspace that the U.S. had ever fought in. The seam between the two cities was the worst of it, because the lethal range of most of the SAMs in both cities overlapped there. We actually had a name for that strip of airspace: Satan's Crack. Our mission comprised of two parts. After air refueling in Hungary, we would drop south and cover a wave of USAF F-16s and Spanish Air Force EF-18s striking a target array just to the west of Novi Sad and Belgrade. Then, we would cover a pair of RAF Tornados, callsign Pistol 11, going after a target in between

Novi sad and Belgrade. This was the part of the mission that was so important.

As amazing as it was that I was sent over to plan my part of the mission in person, I was really surprised to learn that the RAF pilots who would fly the mission were there to plan with me as well. We found a table off to one side and started discussing the mission and figuring out how we would pull it off. For my part, covering the F-16 and EF-18 strike on the front end was no big deal, since they would not fly deep into the threat array around Novi Sad or Belgrade, and they would be on and off target very quickly. But the target for the Tornados presented a really complex problem to solve. To this day I have no idea what that target was because only the Tornado pilots had a "need to know." The reason Tornados were selected for the strike, was that they could carry a type of bomb with good gliding characteristics. The mission required them to fly their gliding bombs into the entrance of a tunnel in the bank of the Danube River. To pull that off they needed to do two things. First, they needed to overfly the tunnel entrance so that they could use their targeting pods and radars to get very precise information for their weapons delivery. Their second requirement was to release their weapons at an exact angle, altitude, and range in relation to the tunnel entrance. That didn't sound too bad at first, but when you laid all of that out on the map you were confronted by a cold reality. The Tornados would have to fly the length of Satan's Crack from west to east to study the target, and then fly back along Satan's Crack in the opposite direction to drop their bombs within delivery parameters!

Flying along Satan's Crack would mean that SAMs could shoot at them from both sides, as well as from in front and behind too. There was a shitload of AAA to deal with as well. It wouldn't be just the 37mm and 57mm stuff, but also the radar directed 100mm and 130mm AAA that could reach up and get you at high altitude. And of course, the route was so close to MiG bases that if the Yugoslavian Air force launched their fighters they would be on the Tornados in seconds after takeoff. No wonder my Ops O wanted our "A-Team" on this mission.

The Tornado pilots and I spent several hours working out different ways to orchestrate the attack. We didn't have to simply plan their strike, we also had to work it in with the F-16 and EF-18 part of the mission. Additionally, we needed deconflict our tactics from other missions happening around the same time nearby. Eventually we agreed on a plan that we felt maximized the chances of the Tornados striking their target and returning alive. My four-ship would air refuel in Hungary and fly south into Yugoslavian airspace. Approaching Novi Sad Cutlass 33 and 34 would split off and set up an orbit just to the east of Novi Sad proper, on the northern edge of Satan's Crack. I would continue with Cutlass 32 and set up an orbit on the southwest side of Novi Sad. From that position I would be able to cover the F-16s and EF-18s as they came in from the west to hit their targets. The F-16s and EF-18s would exit the way they came in. Meanwhile, the Tornados would have refueled over the Adriatic Sea, and would come in from the west, timing their arrival to pass the F-16s and EF-18s as they came off target and headed for home. As the Tornados (Pistol flight) came by, I would swing in behind them, close enough that I could find and attack any SAMs that popped up underneath them, but far enough back to fend-off any MiGs that showed up. Just to the north of us, Cutlass 33 and 34 would be in a perfect position to shoot HARMs at any SAMs, and they could also rush in if we ended up in a big air-to-air battle. Pistol 11 and 12, with my two-ship in trail, would then fly east down Satan's Crack for the Tornado's sensors to pinpoint the target, then turn around and fly back up Satan's Crack to the west and drop their bombs. Assuming everything went well, Pistol would keep heading west back to the Adriatic Sea, and I would turn north and rejoin with Cutlass 33 and 34. From there we would get gas again from a tanker in Hungary and go back to Aviano. By the time I left the MPC that night I was pretty excited about the mission. It would be a tremendous challenge to pull this off. It would push not only my stick and rudder skills to the limit, but also my ability to quarterback in the heat of combat. I was stoked!

The next evening, I met up with the rest of Cutlass flight, and we had dinner at the hotel we were staying in before heading down to Aviano. Once at the squadron I checked in with Pistol flight over the phone while the rest of Cutlass flight retrieved updates on the weather (clear over the target area), visited Intelligence for a threat update, checked on the status of our jets, and the dozens of other little tasks needed before we briefed the mission. When we closed the door to the briefing room and I handed out the maps and other information for the mission, the room got really quiet. It was the first time the other three pilots had seen what I had planned, and they were soaking it all in. My wingman broke the silence. "I thought we were flying Force Protection tonight, not a suicide mission?" That broke the tension, and we all had a good laugh over that…even though it might be true!

Everything went smoothly from there, and before long we found ourselves at our jets getting ready to fly. Preparing to fly a combat mission is quite a chore, and it is probably worth taking a moment to explain all of the things that happen between when the briefing ends and a pilot puts his foot on the ladder and climbs into the jet. Usually, you only have about 10 minutes after the combat brief before it's time to walk out of the squadron. The restroom is usually the first stop after the briefing, and then everyone meets at the Operations Desk. At the Operations Desk we would get last minute updates on the weather and any important intelligence. A senior pilot, called the Top 3, is on duty there as well, and would cover any other information we need to know about the airfield or the status of our airplanes. Finally, we were assigned the tail numbers for the jets we will be flying along with their parking locations on the airfield. Unless there were questions, everyone piled into a crew van which took us over to the Life Support building. Life Support is where all of the flying gear is stowed and maintained between flights. Each pilot had their own locker containing their parachute harness, helmet, G-Suit, survival vest, and assorted other gear. When flying combat, the first thing you would do to get dressed for flying was "sanitizing" your flight suit.

This entailed removing all of your patches and emptying your pockets of any personal effects. You do this so that if you get shot down and captured, the enemy wouldn't learn anything personal about who you are or what squadron you fly with. All you carry are your dog-tags and your military ID card, the latter of which is also your Geneva Convention ID card that identifies you as a military member eligible for treatment in accordance with the Geneva Convention for the treatment of prisoners of war. Most guys also carried some cash too, so they could get something to eat if they diverted to another airbase instead of returning to Aviano. Next you zipped on your G-Suit, put on your survival vest, strapped your parachute harness over all that, and tossed your helmet in your helmet bag. Lastly, you signed for your pistol and spare magazines, loaded a live round in the chamber, and snapped it into the holster on your survival vest. Most guys also threw a few water bottles into their helmet bag too, and maybe even some food. Having plenty of water was often more important that you might think. If things changed or something went wrong once airborne, you could easily wind up flying for six to nine hours before having a chance to get a drink again. As much of a bulky pain in the ass it was to carry and stow in the cramped confines of an F-16 cockpit, I always packed 10 water bottles. Once everyone was dressed, we would all climb back into the crew van and go out to the jets.

Arriving at the jet, the crew chief would greet you with a salute, and hand you the maintenance forms for that specific airplane. It was pretty common for your crew chief and his assistant to be sweaty, dirty, and tired from the hours of work it took to get their jet ready to fly. These young men and women took a lot of pride in giving the pilot a jet that was ready to fly, and I never in more than 20 years in the Air Force ever saw anything less that true professionalism from the crew chief cadre.

After reviewing the maintenance forms and signing for the jet, the next task was to start the pre-flight. While the crew chief would start pulling different safety pins and covers from the airplane, I would pop open a little fuselage panel beneath the cockpit and insert my video tapes into the recording machines. After that, I would climb up the boarding ladder and unpack my stuff. I would sit on the rail of the cockpit and lean in, putting the Data Transfer Cartridge into its socket, sitting my helmet on the far canopy rail, and then set all of the switches in the cockpit to the proper positions for engine start. Climbing back down, I would then do a walkaround of the whole jet, checking various gauges, pins, panels, and settings to make sure the jet was ready to fly. That accomplished, it was normally time to climb into the cockpit and get strapped in.

On this night, however, we were running a few minutes ahead of schedule. I took advantage of this to show the crew chiefs the map of where I was taking their jet and explained what I was going to do with it. I always did this when I had the time, because I wanted the crew chiefs to know that all of their hard work turning wrenches in the heat and cold was important. I think they always enjoyed it too and added to their sense of pride in helping accomplish the mission. On missions where I shot HARMs or dropped bombs, I always took a few minutes after the flight to tell them all about it.

After getting into the jet everything went smoothly. Engine start, taxi, takeoff, and the flight to our tanker in Hungary was completely uneventful. Right on schedule, we fenced in and contacted the NATO AWACS to let them know we were on station. The AWACS acknowledged and let me know that all of our strikers were on time as well. As we approached Novi Sad from the north I was once again amazed that the Yugoslavians left their city lights on when they could see us coming. I didn't even need to use my navigation equipment to find my way, which freed me up to spend more time looking for SAMs, AAA, and MiGs.

At the appointed time, Cutlass 33 and 34 peeled off for their orbit, and as my wingman and I crossed over the city the AAA started hunting about the sky with a few streams of tracer fire.

Avoiding the worst of it, we had soon picked our way across Novi Sad and turned into our orbit just a few minutes ahead of the F-16s and EF-18s beginning their bombing runs. Studying my sensors, I could see that the Yugoslavian early warning radars were up in full force, but I decided to not shoot them, saving my HARMs for the SAMs that I knew we would have to deal with later. Looking off in the distance I could see some AAA over southern Belgrade, but that was no factor to us. After a couple of trips around our orbit, I heard the F-16s and EF-18s call that they were on their target runs. The Yugoslavians didn't fire so much as a single round of AAA at these guys, and soon they were off target. As I watched them exit to the west on my air-to-air radar, I saw Pistol flight inbound. Everything was going as planned so far, and I was optimistic that they might stay that way. With all of the early warning radars banging away, I was sure the Yugoslavians could see that we really didn't have many jets in the sky tonight, and maybe that would lull them into not getting too aggressive with the SAMs as we flew down

Satan's Crack. Maybe they would see that we weren't heading for downtown Novi Sad or Belgrade, and decide we weren't worth a big effort?

As Pistol flight flew by, invisible in the night sky, I swung around and saddled up behind the Tornados like we had planned. I checked my young wingman, and he was in perfect formation, about three miles behind, offset to one side, and stacked 2,000 feet higher than me. Looking out the front of my cockpit, both Novi Sad and Belgrade were a sea of lights, and the Danube River looked like a dark ribbon weaving through them. As we approached the edge of that sea of lights, the AAA started to increase in intensity, and my adrenaline began flowing. At first it started out as a few strings of 37mm and 57mm tracking up into the night sky, bursting red and orange below our altitude. Then I started noticing the white flashes near us at the same altitude. These were the 100mm and 130mm AAA shells. Their shells were programmed to detonate at a predetermined altitude...our altitude...which had probably been initially provided by the early warning radars we had seen earlier. Very quickly I began to see radars popping up around us that would provide detailed tracking information to the AAA, making their aim more precise. To complicate their aim, I began the familiar dance of weaving my jet left and right while changing my altitude up and down slightly. I also turned on my electronic countermeasures pod, and at random intervals deploying a few bundles of radar-spoofing chaff. While I made sure to watch my radar and other sensors closely, I spent most of my time looking outside the cockpit to dodge the ever-increasing volume AAA, and the SAMs that I knew would follow.

As we passed by the target eastbound in trail of Pistol, things began to really heat up. Cutlass 33 and 34 started to call out a variety of SAM radars popping up around us, and my RWR scope began to chirp and beep as different threat radars looked at my jet. And then all hell broke loose. My RWR scope screamed as it detected an SA-6 launch at my F-16, and looking low and right, I saw the missile already airborne and closing on my jet! I shoved the throttle into full afterburner and made a quick "Cutlass 31 defensive" call over the radio. The missile was already too close to my jet to try my normal guidance-disrupting tactics. My only option was a violent maneuver that I hoped would generate enough distance between me and the missile that it would miss at the last instant. Never taking my eyes off of the missile, I muscled the stick back and to the right, executing a high-G barrel roll into and over the approaching SAM. As I was upside down in my roll the missile passed underneath me, close, but not to close that its proximity fuse triggered a detonation. But in that moment, I didn't have time to celebrate, as what I saw beneath me captured my full attention. Upside down, looking out the top of my canopy the city lights were bright and vividly crisp. The sky in between me and the lights was absolutely alive with hundreds of rounds of AAA tracers arcing upwards or exploding. And rising up directly toward me was another SA-6 missile! I don't recall it, but Cutlass 33 told me later that I keyed the microphone and called "Another 6" over the radio. What I do remember was pulling the nose straight down to dodge this missile, that I was positive was about to hit me from below. Halfway through this high-G pull the missile flashed by my jet, passing close aboard, over the top of my jet from front to back, and detonating just past me. I didn't have any time to decide if I had actually been hit or not. I was now diving straight down into that beehive of AAA fire.

Unlike my last night of close calls over Belgrade, this time I had plenty of speed, so my F-16 was wonderfully maneuverable. With the afterburner still cooking, I rolled and twisted my jet around the curtains of AAA fire, looking for more SAM launches and clawing skyward to get above the worst of the shooting. I saw a missile streaking in high from the north. It was a HARM, and I remembered hearing Cutlass 34 firing it at one of the SA-6s that shot at me earlier. I saw several other SAM launches, but they weren't aimed at me, Cutlass 32, or Pistol flight. Rolling out and leveling off at 25,000 feet, I canceled the afterburner and turned my attention to making sure the Tornados, still out in front of me, were okay. I could see them on my radar, and I hadn't heard any radio calls from them, so I knew they were holding their own.

At that moment Pistol flight reached the far eastern end of their route and executed a 180 degree turn to headed back down Satan's Crack on their bombing run. I followed them around the turn, and almost immediately began to dodge a new wave of 100mm and 130mm AAA fire that was too close for comfort. I also detected several SAM radars, but none of them were in a position to threaten us, so I left them alone. I was still jinking all over the sky, and looking for SAM launches and MiGs, when Pistol called that their weapons were away. Looking forward out of the cockpit and around the nose of my jet, I could see Pistol flight light their afterburners, accelerating and climbing. I turned my attention to the target area, and in a few moments, saw the ghostly yellow-orange flash of an explosion along the bank of the Danube…mission accomplished! I continued to trail the Tornados until they were clear of Satan's Crack and heading westbound for home. Our charges safely away, I turned north, and called Cutlass 33 to let him know I was northbound and for him to rejoin on me enroute to the tanker. As we left Novi Sad and Belgrade behind us, I could see the AAA slow and then stop all together.

It was then that I noticed my Air-to-Air Tactical Air Navigation (TACAN) display was reading zero distance, and not locked onto any signal. The Air-to-Air TACAN is a piece of equipment that allows two airplanes to measure the distance between them. It was a very useful when we flew at night since it helped us keep track of where we were in relation to each other. Normally I would expect to see between three and five miles between me and my wingman. At first, I wasn't too concerned at seeing the Air-to-Air TACAN had broken lock. That happened sometimes. But then I thought about all of the AAA and SAMs and realized I hadn't heard anything from my young wingman, Cutlass 32, since before we rejoined on Pistol flight. I keyed the microphone and made a call over our inter-flight radio frequency, "Cutlass 31 check." There was a long pause, and then I heard "Three" crisply followed by "Four." This meant that Cutlass 33 and 34 had heard me, but Cutlass 32 didn't. I tried again. "Cutlass 31 check." There was another long pause, before Cutlass 33 and 34 acknowledged once more. Now I was starting to get worried. Where was my wingman? Had he been shot down and nobody saw it? I tried the inter-flight radio again. "Cutlass 33, this is Cutlas 31. Are you hearing Cutlass 32?" My Ops O quickly replied, "Negative." I knew that Cutlass 33 and 34 were about 15 miles behind me, so I asked another question. "Cutlass 33, do you see Cutlass 32 on your radar anywhere?" I was hoping that Cutlass 32 had simply lost his radio and couldn't answer but was still in formation behind me where he should have been. After a short delay, Cutlass 33 replied, "I don't see him anywhere." From the tone of Cutlass 33's voice, I could tell he shared my concerns.

I had one more idea. I would try and reach Cutlass 32 on the radio we kept tuned to the strike frequency. "Cutlass 32, this is Cutlass 31 on strike, do you copy?" All I heard was silence. As a last option, I decided to call the AWACS and see if he knew where my wingman was. "Magic, this is Cutlass 31. Do you have a position for Cutlass 32?" There was a long pause, and then the reply, "Cutlass 31, Magic, negative contact with Cutlass 32." There it was.

My wingman was gone. My heart sank, but I quickly put that feeling away. I had work to do if there was any chance of rescuing him.

I took a deep breath before speaking. "Magic, Cutlass 31. My wingman Cutlass 32 is down somewhere in the target area. Cutlass 31, 33 and 34 will go to our tanker and get gas while you start the rescue effort. We will return in 15 minutes, and I will be the on-scene commander until the CSAR forces arrive." Magic acknowledged, and I started to think about what I needed to do next. If Cutlass 32 was able to eject, the odds are he was captured as soon as he landed in his parachute since we had been fighting over a heavily populated area. I couldn't imagine any scenario where you parachute into a heavily populated city in the middle of a war, and nobody notices you. I dropped my oxygen mask and pulled out one of the 10 water bottles I had brought with me and took a long drink. It looked like I had a long night in front of me, and I would probably empty them all before it was over.

There is an old saying, "How do you eat an Elephant? One bite at a time." I had a lot to do as the on-scene mission commander for the rescue, but the "first bite of the Elephant" was to efficiently get my flight to the tanker to get gas. I stowed the water bottle and prepared to fence out as the we approached the border with Hungary. I didn't need to say anything on the radio, I knew that as we crossed the border Cutlass 33 and 34 would fence out automatically; it was standard squadron procedure. Fencing out was the opposite of fencing in. Instead of turning all of your weapons on and getting ready to fight, you safe everything up and turn back into a normal airplane again. The last thing on my fence out checklist was to turn my exterior lights back on. As I reached for the switches, I was shocked to see another airplane turn on its lights right in front of me! This airplane was not more than a few hundred feet in front of me but was flying about a thousand feet higher. I paused my hand on my light switches and stared. The lights were in the same pattern as an F-16's light. The most wonderful feeling of relief washed over me. I was pretty sure I had just found my lost wingman!

I turned on my lights and called Cutlass 33 and 34 and let them know that there was now another F-16 flying above me headed to our tanker. I also let Magic know, and to have the CSAR forces standby until I could verify this mystery F-16 was indeed Cutlass 32. I stayed below and behind him, and sure enough he flew right to our air refueling tanker. As we approached the tanker, I switched us from the strike frequency to the air refueling frequency.

Keying the radio on that frequency I called out, "Cutlass 31 check." I was immediately answered by "Two, Three, Four" as Cutlass 32, 33, and 34 acknowledged. I laughed out loud! I called Cutlass 32 and asked him to check that he had his other radio tuned to our inter-flight frequency. In moments he checked in on that radio frequency as well. After a quick call back to the AWACS to let him know Cutlass 32 was safe, we topped off our gas tanks and flew back to Aviano. After landing we got together and figured out what happened.

Cutlass 32 was in perfect formation with me as we began flying into Satan's Crack. And like me, he was having a devil of a time dodging AAA and looking out for SAMs. Somehow, he managed to switch both of his radios to other frequencies, turn off his Air-to-Air TACAN, and turn his IFF transponder to standby...all without noticing that he had done it. To this day I try and construct scenarios of how you could actually do all of those things accidentally, because they are not easy to accidentally do in an F-16. We didn't believe him at first, so we watched the video tapes from his jet that recorded his avionics settings. And sure enough, over the span of a couple of minutes, you could see those frequencies change and the other equipment get turned off. I guess it just goes to show you, that crazy things really can happen in the heat of battle. It was also a cautionary tale to never underestimate the ability of a young fighter pilot to fuck things up. So, by the time I was dodging those SA-6s, Cutlass 32 was sitting happily in his cockpit, blissful in the silence of a quiet radio, weaving through the AAA. He saw the SAMs, but didn't know I was dodging them, so he kept flying straight ahead. By the time I finished defeating the SAMs and was saddled back up behind the Tornados, he was directly above me. He did know that he had lost track of me and called me on the radio (over the wrong frequency), but when I didn't answer he decided to just stay behind Pistol flight and follow them. He continued to stay at his altitude and fly the flight-planned route. Apparently, that kept him pretty much with me all the way, even though he had no idea where I was. When he reached the last navigation point on the Tornado's route of flight, he decided to just go to the tanker in Hungary and look for me there. As it turned out, that was a great plan. And made for an even better story with a happy ending!

Chapter 7: Gutsiest move I ever saw, Maverick

This particular story takes a little setup, so bear with me! When a squadron goes to war there are a lot of things that need to be done every day to make that work. During peacetime, all fighter pilots have what are called "additional duties" in the squadron. Yes, your primary job is to fly and fight, but the squadron needs pilots to take on the jobs that only other pilots can manage due to the background and experience they require. Squadrons need training officers to manager all of the ground and flight training pilots require. There are pilots who collect all of the different meeting schedules, flight schedules, training requirements, etc., and build a weekly flying schedule that integrates all of that. There is the Standards and Evaluations office, which manages and gives check-rides. It takes dozens of more jobs than these to make a squadron work, so every pilot has an "additional duty" to go along with their flying. Some pilots have more than one.

Day to day during peacetime, each pilot is also responsible for planning their own missions. This is important, because flying an F-16 mission does not just happen on its own. You have figure out where you are going, how to get there, and what you are supposed to do when you arrive. The process for flying a low-level navigation route to a pop-attack, with 2,000lbs. bombs, is very different from an air-to-air only mission, or a Force Protection mission. And just about every day the training mission is different from the one you flew the day before. So, when you see your name on the flight schedule, you get together with your flight mates (usually the day before) and plan the mission. You look at what kind of mission it is, what ordnance and external stores the jets are carrying, what airspace you have available, what training items each flight member needs to complete, etc. Then the leader of the flight divides up all of the planning tasks that are needed, be it drawing maps, calculating weapon delivery parameters, or building the Data Transfer Cartridge (DTC) file that gets loaded into the jet with all of the information the F-16's computers will need. After all of that is done, the pilots who are flying together assemble in a briefing room and go over the upcoming flight in great detail. Only when all of that is accomplished do you go fly. After the flight, the mission is debriefed and reconstructed in even greater detail. It's no exaggeration that a one-hour flight can take two or three hours to debrief. During the debrief everything is dissected under a microscope, and everything that is both good and bad noted. There is no rank in the debrief, either; a General receives the same critique as a new Lieutenant. And these debriefs can be brutal…if you fuck something up you will take your lumps for it. This no-holds-barred honest level of critique is why the U.S. military has the best fighter pilots in the world. The fighter pilot culture is dedicated to perfection. Incidentally, some folks who are fine pilots leave the fighter community because they can't separate this professional honesty from a personal attack. They will argue and make excuses, and the fighter pilot community has no tolerance for that. For new, young pilots learning to fly fighters, the pressure of the debrief can be a rude awakening. No kidding, when I was an instructor at Luke AFB teaching new pilots to fly the F-16, I saw or made grown men and women cry during the debrief quite a few times.

But when a squadron goes to war, the pace of operations changes. All of the normal squadron additional duties still need to be accomplished in spite of what can be a brutal pace of flight operations. To make it even more complex, you aren't flying with only the guys in your squadron. You have teammates flying different kinds of airplanes and from different countries to work with. Oh, and it is a real combat mission, where people are going to get killed and things blown up, so you can't stop and reset the scenario if the plan gets screwed up. Combined with the fact that you don't know what kind of mission will be flown until a few hours before takeoff, it means that the pilots flying the missions rarely have the time do the planning for themselves. All of this complexity means that a more efficient way to plan is to have a Mission Planning Cell (MPC). A wartime MPC is comprised of pilots from all of the air forces and airframes that are fighting together. Since Operation Allied Force was a NATO operation, we had a huge number of people in the MPC.

Every day, as an additional duty, we scheduled four pilots as our squadron's contribution to the MPC. This always consisted of one highly experienced guy to make the final decision on the best tactics, techniques, and procedures for the different missions that would be flown in the upcoming 24 hours. These four pilots would do the planning for all of the squadron's missions in the coming day, coordinated with the MPC representatives from the other NATO air forces. We would work out all of the details, from takeoff times to times over target, draw the maps, calculate weapons delivery data, load the DTCs, etc. The squadron's MPC team would then provide all of this information in a package to the guys when they showed up to fly. It was a shit-ton of work done under immense time constraints. So, all of that said, at least once a week you would be scheduled to work a shift in the MPC. And that was how one of my most interesting missions started.

I was leading our squadron's MPC team one night. It was around 7pm when the Air Tasking Order (ATO) came in. The ATO is the master schedule of the air war covering the next 24 hours. The ATO was built by the NATO Combined Air Operations Center (CAOC) staff in Vincenza, Italy. When the ATO comes in, a frenzy of activity starts in the MPC. The first thing that happens is you have to search the hundreds of pages in the ATO for those missions that apply to your squadron. This is called breaking out the fragment, or the "frag" for short. On a given night the frag generally had around six missions for our squadron, each requiring 4 airplanes and pilots. We would also dig through the frag and figure out what other squadrons our guys would be working with, and then begin coordinating with those units to build a plan based on the mission. That particular evening, one of the missions called for our squadron to provide a flight of four F-16CJs to protect a flight of four French Air Force Mirage 2000s striking a target next to the international airport in Pristina, the capital city of Kosovo.

I could write a lot about how painful it was to work with the French Air Force, not only during Operation Allied Force, but during Operations Desert Shield and Desert Storm, and also over the Iraqi No-Fly zones. Perhaps I can best describe the French fighter pilots as needy, untrustworthy, and honestly not that good at what they do. Regardless of that, it was my job to plan the best mission I could with them, so that's what I got to work on.

The Mirages were going to strike their targets with Laser Guided Bombs (LGBs) from medium altitude, which would keep them well above most SAMs and AAA. Pristina was not that "hot" of an area to work over anyway, as any real threats and difficult targets had been eliminated in the opening days of the air campaign. When I approached the French pilot planning their strike, I thought it would be a pretty straightforward and easy mission. The error in my thinking quickly became apparent, as the French pilot, in an almost comical stereotype of a panicked Frenchman, began stabbing the map with his finger. "There is a SAM at the Pristina airfield, you must destroy it before we can attack!" he practically shouted. I looked at him, and then at the map he had spread out. It showed a single SA-6 at the airport. I looked back and forth between the map and the French pilot a few times, gathering my thoughts into a diplomatic reply. "Well," I began, "while an SA-6 is certainly a dangerous SAM, I don't think this one represents much of a threat for you. First, you are at medium altitude and are dropping LGBs, which should keep you at the edge of his range, if you enter it at all. Secondly, that particular SA-6 hasn't been seen around Pristina for weeks and given that SA-6s are highly mobile that probably means it moved away somewhere else. And finally, you will have us there to target it if it pops up. So, I think you will be just fine." The Frenchman, however, remained unconvinced. Tapping the map again for emphasis, he replied, "We can only strike this target, then, if you shoot HARMs at it every 30 seconds that we are on our target run and until we are clear of the area." I started to laugh, but he was serious. I shook my head no. To begin with, HARMs aren't belt-fed like the bullets in a machinegun. A flight of four F-16CJs only carries eight HARMS, and he was asking for us to shoot 10 of them, so his plan was a non-starter to begin with. It was also a dumb idea, since SA-6s are mobile, and it had been weeks since it was last parked there. Unless we knew the location of the SA-6 there was virtually no chance of our hitting it by randomly shooting HARMs at the ground. If HARMs had nuclear warheads that might have worked…but they don't.

I tried explaining all of this to him, and I even tried to shame him by pointing out that just a few nights ago my wingman and I trolled back and forth over Belgrade getting the shit shot out of us, and that a single SA-6 (even if it was there and feeling froggy) wasn't that hard to deal with. But he still remained stubbornly demanding. I wasn't really pissed at his insistence…I almost expected stuff like this when flying with the French…it was just a huge amount of drama, when we had a lot of work to do getting all the night's mission planning done. "How about this," I offered, "We will shoot one HARM so that it arrives where that SA-6 is on your map at the time you are closest to it? And we will be right behind you with more HARMs in case he decides to start shooting at you. Does that work for you?" The French pilot didn't even hesitate, "No!" he replied, and stomped off. I watched him weave his way through the chaos of the MPC toward the French Air Force's private office. I shrugged, "Well, at least I can get on with planning the other missions." I thought that was pretty much the end of it, and that he would come back in a while and take my offer. Boy was I wrong.

Apparently, I really whipped the French into a lather. I will never know what went down behind their closed doors, but a few hours later the phone rang, and it was for me. It was our F-16CJ representative at the NATO CAOC in Vincenza. The French government had put its collective foot down and said that the French Air Force wouldn't fly near Pristina until that SA-6 was destroyed. To accomplish that, NATO scrubbed the original mission to Pristina, rolled our F-16CJs to cover another tasking, and added a new mission to destroy the SA-6. This kind of mission was called Destruction of Enemy Air Defenses, more commonly called DEAD (pronounced "Deed"). I took down all of the notes about the mission and called back to the squadron to let them know. Changing the original mission wasn't a big deal, but we didn't have any spare pilots to cover the new DEAD mission. I looked at my watch and realized that me and the three guys I was working the MPC with would have sufficient time to get some sleep and take the mission, and we would have the added benefit of planning it ourselves. I suggested this to the squadron scheduler and leadership over the phone, and they agreed. The mission was mine!

Our DEAD mission consisted of my four-ship of F-16CJs (callsign Bobcat 81) and two F-15E Strike Eagles (callsign Dagger 41), and a couple of tankers for refueling. I was designated to be the Mission Commander, so after we finished all of the other MPC work, we sat down with the F-15E MPC team and developed a plan. DEAD missions come in two basic flavors. The first are ones where the SAMs are at a fixed, known location. The second type of DEAD mission involved going after SAMs that were mobile and their locations not known. This mission fit into the second category. We needed a plan that was very flexible, especially given that our best intelligence told us that the SA-6 wasn't at the Pristina airport any longer.

Acting on the idea that the SA-6 could still be nearby, even if it wasn't on the airfield, we developed a plan to search the whole area around the city. It was really a simple plan, too. Dagger flight was equipped with NVGs and targeting pods, as well as a shitload of LGBs. They would stay about 25 miles away from Pristina, and wait for me to locate the SAM. I would do this by making him shoot at me. Using our specialized equipment, the F-16CJs would be able to get pretty good coordinates on the SAM if he turned his radar on, and then we could pass those to Dagger to bomb. Even if we didn't, get coordinates for the SAM, Dagger's NVGs and targeting pods would allow them to actually see the spot on the ground from which the SAM launched, and then attack it with their LGBs. Bobcat 83 and 84 would stay with Dagger flight to protect them if they should find themselves the target of any SAMs. My wingman, Bobcat 82, would fly a wide circle around Pristina, ready to turn in and shoot a HARM at the SA-6 (or any other radar-guided SAM) if it happened to pop up near my location. And me? My job was to fly back and forth over Pristina in an effort to get the SAM to shoot at me and give his position away in the process!

After planning, we all went and got breakfast, and then drove back to the hotel for some sleep before it was time to go fly the mission. Fortunately, we weren't scheduled to take off until pretty late that night, so we all got some good rest, and met up for dinner before reporting in at the squadron. On the way to the briefing with Dagger flight, we stopped by our intelligence shop. I had put in a special request for any intelligence information on the SA-6 in question, as well as any other SAMs that might be in the area. As expected, that SA-6 hadn't been seen in weeks, and no other radar-guided SAMs has been seen around

Pristina recently either. We were warned, however, that we could expect the Yugoslavians to have an arsenal of AAA and heat-seeking man-portable air defense system (MANPADS) SAMs in the area, many comparable to the U.S.-manufactured Stinger missile. Normally we didn't sweat MANPADS too much because we stayed above their maximum altitude. But you still had to be careful because of the high mountains around the area. It would be easy for a bad guy to climb up on one and take a shot. The other thing about MANPADS was that because they are heat-seeking, you don't get any radar indications to let you know you are being tracked and shot at. You have to see the missile launch and react on that indication alone. After our intelligence briefing, we got an update on the weather (clear skies for a change), and then met up with Dagger flight to discuss the mission. From there we held a quick briefing with just our four-ship before stepping out to the jets. Since we four had planned the whole thing, no one had any doubt about their role in the upcoming mission. For once we were running a little bit ahead of schedule, which was nice, and nobody even had the slightest problem with their airplane. It almost seemed too easy as we taxied out and took off.

It was a beautiful night to fly. There were no clouds, and you could see forever, with the stars visible in amazing abundance! The flight down the eastern coast of Italy was equally peaceful, and before long we turned across the Adriatic Sea and rendezvoused with our tanker. Dagger flight was on time as well, and after checking in with the AWACS, we fenced in. It was game time.

Departing Albanian airspace, we split into the formations we would fly, with Bobcat 83 and 84 peeling off to stay with Dagger, and my wingman and I proceeding directly to Pristina. We were the only NATO airplanes over Kosovo that night, and when I began to pick up indications that the Yugoslavian early warning radars were looking at us, I hoped they would take the bait and play. Early warning radars aren't directly hooked up to SAMs. Instead, they provided information that allowed the Yugoslavian air defense command and control structure to organize their response to our presence. Hopefully that would mean telling their SAMs in the area to engage us.

As we approached Pristina Bobcat 82 angled away from me to set up his arc nearby. I looked out from the cockpit and spotted Pristina's airport quite easily, because once again the Yugoslavians had left the city's lights on. I decided to cross the airport at 25,000 feet and 450 knots, which would give me plenty of altitude and airspeed to dodge any missiles that came my way. As I cruised across the city, I gently weaved and rolled my jet from side to side so I could watch the ground below for missile launches. Not seeing anything of interest during my first transit across Pristina, I decided to set up an orbit over the airport itself. After about five minutes of making left-hand turns, with nothing happening on the ground below, I began to wonder if the whole night would be a waste of time.

As I stared down at the city, I decided that maybe I needed to make myself into a juicier target by doing something we called "stimulating the threat." "Perhaps," I thought to myself, "they know that with my airplane at 25,000 feet they have a lower probability of hitting me, so they are holding their fire. Maybe if I fly a little lower, I will look like a more lucrative target for them. One that they can't resist!" With my decision made, I let my flight mates know that I was going to try a lower altitude. I cracked the throttle back, and let the jet descend while maintaining 450 knots. If "stimulating the threat" by flying lower worked, I still wanted plenty of airspeed to maneuver.

Leveling off at 20,000 feet things started to happen. The first indication of an enemy reaction was some AAA fire out of the airport area. It was 37mm and 57mm stuff and given the lack of indications of a targeting radar, I knew they were simply shooting into the night sky at the sound of my engine. None of it came anywhere close to my jet, and the shells were air-bursting comfortably beneath me. It was actually pretty cool to watch, kind of like watching a firework display from above. After a couple of more circles over the airport area I still hadn't seen any indications of an SA-6 radar, or any other SAM or AAA radars for that matter. So, I decided to expand my orbit and basically wander around over all of the city. This resulted in more sporadic AAA being randomly fired into the sky. It was certainly nothing like the volume of fire I had seen over Belgrade on other nights, but it was better than my tactics drawing no reaction at all! We had enough gas to stay over Pristina for about 45 minutes before going back to the tanker for more gas, and before I knew it that time had come.

As we headed back to the tanker over Albania, I unhooked my oxygen mask and pulled out a bottle of water. As I sipped it, I thought about the tactical problem that needed solving. It was apparent that based on what I had presented to the Yugoslavians, randomly fired AAA was all they were offering in response. What we were doing, and how we were flying, was definitely different than the way missions normally looked over Kosovo. Perhaps they smelled that I was setting a trap for them if they activated one of their radars? The question I needed to answer was, "What do I need to put on the table to make them play? What could I do that would make them salivate so badly that they would take the risk of getting blown up themselves in the process of shooting at me?" As we neared the tanker, I stowed my water bottle, reconnected my oxygen mask, and focused on rejoining with the tanker and getting our gas. I had a few ideas and let them stew in my brain while we all took gas. By the time we had departed the tanker I had a new plan and briefed my wingmen and Dagger flight on what I was going to do.

I decided that the Yugoslavians must have recognized that I was bait for a trap, and they weren't going to bite. But they did shoot enthusiastically at me with 37mm and 57mm AAA. My guess was that despite their centralized command and control seeing the trap I was setting, some of the troops on the ground were either not getting the word to sit on their hands, or they just couldn't help themselves and had to pop off a few rounds at the jet flying around over their city. Perhaps they saw it as an insult? After all, Yugoslavians are a known to be a proud people, and it had to gall them to know I was flying around their country with impunity and distain. I recalled that Pappy Boyington had taken advantage of Japanese honor and pride back in World War II on one mission, to great success. He had led his squadron of Corsairs over the Japanese airbase on the island of Rabaul, hoping that their presence would bring about a big air battle. The Japanese, however, didn't fall for it. Knowing that the Japanese pilots valued their honor to an extreme, and also knowing that the Japanese monitored the American radio frequencies, Pappy began taunting the Japanese over the radio. He even told them exactly where he and his squadron were located and at what altitude they were flying, although he lied about the altitude to give himself a tactical height advantage. His insults about their Emperor and calling them cowards did the trick, and the Japanese pilots launched in mass, only to be shot down in mass as well. Maybe I could pull something off like that? The Yugoslavian gunners and SAM operators wouldn't be listening directly to our radios, so taunting them that way was unlikely to work. But maybe I could taunt them with the way I flew? Maybe I could make myself into such a juicy a target that they would know I was basically giving them "the finger," and daring them to do something about it? It was worth a try!

Approaching Pristina for the second time that night, we set things up exactly as before, with Dagger ready to put bombs on any SAM sites we could find, and my wingmen set up to cover everyone with HARM shots if needed. Driving straight toward the middle of the city at 20,000 and 450 knots I could see that the Yugoslavians knew we were back, as the light and sporadic AAA started up again. After a couple of orbits around the city, and not drawing anything more than the previous level of AAA, I decided it was time to send my first insulting message. I turned on all of my exterior lights.

I know that might not sound like much but recall that normally we cruise around blacked-out so that the bad guys can't see where we are. That means they have to use radar to find and engage us, which opens them up to our locating and attacking them. The AAA I had seen over Pristina so far that night was obviously not aimed at anything other than the sound of my engine and was inaccurate as a result. But with my exterior lights blinking and flashing on their brightest setting, anyone looking up at the clear night sky could easily see me. That would allow them to aim their AAA without using radar. Additionally, many different SAM systems have the ability to track and engage a target optically, including the SA-6. Maybe if the SA-6 didn't have to turn on his radar he would take a shot at me, and in the process reveal his location to the F-15Es, who were searching for him visually with their targeting pods and night vison goggles.

The effect of turning on my exterior lights was noticeable, with the AAA increasing appreciably in both volume and accuracy. Even though I was above the effective range of most of what they were shooting at me, I still kept weaving my jet back and forth. I wasn't too concerned about the AAA, but I had to respect it, since the Yugoslavians did have bigger AAA all the way up to 100mm…and that could definitely reach up and tap me on the shoulder if I wasn't careful! Plus, I needed to keep a really good lookout for SAM launches, since if my plan worked and they engaged me optically, I would have no indications on my RWR to alert me that I was being shot at. After several minutes I was beginning to get discouraged again. They were shooting AAA at me with great enthusiasm, but that was all. I was pondering my next move when a flash of light on the ground caught my attention: a SAM launch!

I could tell right away that it was a MANPADS SAM launch, which meant it was a heat-seeking missile. MANPADS SAM launches have a very unique look with the way the missiles kind of wobbles as it flies up. They are fast too, like a bottle-rocket. Most MANPADS can't reach up and touch an F-16 effectively at 20,000, and with my high airspeed I was in a good position to defeat the SAM with flares and some maneuvering. I called "MANPAD launch, left and low" over the radio so that my HARM-toting wingmen wouldn't take a shot and spoil my plan by spooking the bad guys. I grinned in my oxygen mask as I watched the SAM try and match my maneuvers, run out of speed, and fall away well beneath my jet. Weighing how the missile looked and the relative ease with which I defeated it, I decided it must have been an SA-7, which is an older MANPAD system. I very deliberately cautioned myself to not let that little tidbit cause me to relax, because the next SAM could be a more advanced MANPAD, or even the SA-6 we were hunting. Complacency and overconfidence are killers, in combat or peacetime. I was really happy with this change of events. As I wandered around the sky the AAA kept following me, and the SA-7 launch was an encouraging sign that more might be coming my way.

This went on for quite some time, with another MANPAD launch that was even less close than the first one. Glancing at my fuel gauge and my watch, I realized that we didn't have a lot of time left on station before we would need to return to the tanker and fly home to Aviano. I was pleased that my turning the lights on had resulted in some good action, but I needed to up the ante if we had a chance to make anything more happen with the time and gas we had remaining. Over the radio I let everyone know that our time on station was almost up, and that I was going to push things a little farther, and to keep their collective eyes peeled. This would be our last chance to get the SA-6 we came for. My plan now was to fly my last few orbits lower and slower to make myself such a tempting target that no self-respecting SA-6 crew could pass up the chance to take a shot. In a few moments I leveled off at 15,000 feet and had steadied my jet at 350 knots. Tactically speaking, this was relatively low and slow, and made me a very tempting target.

At this speed and altitude, I was in the heart of the engagement envelope for the SA-6 (and any other radar-guided SAM, for that matter), as well as within range of the AAA and the MANPADs. The Yugoslavians didn't seem to notice the difference, but I sure did! Now I was having to aggressively maneuver to avoid the AAA fire, jinking my jet left and right and up and down as the strings of tracer fire chased me across the sky. But as long as I kept my head on a swivel, I had no real trouble staying clear of it.

Time was running out, and I was down to my last pass over the city. I decided it was time for the last thing I could think of to draw a reaction. I decided to drop a few flares and tap my afterburner in addition to having my exterior lights flashing. The noise from the afterburner might be noticeable on the ground, and the flares and the afterburner plume would basically leave a flaming dotted line across the sky pointing to my jet. If that didn't get the SA-6 to shoot, then nothing would.

The effect was immediate and impressive! I had no sooner tapped the afterburner and pumped out some flares when the somewhat lazily aimed AAA seemed to snap at my jet all at once. I broke hard to the left and slammed the throttle into full afterburner to avoid it. At that altitude the F-16 has amazing performance, even with all of the HARMS, gas tanks and air-to-air missiles hanging under it. The strings of AAA fire had just passed behind and to the side of my jet when I saw another MANPADS launch beneath me. Knowing that the afterburner plume acted like a magnet for the heat-seeking sensor on the missile, I snapped the throttle back to idle and deployed a string of flares to decoy the missile. At the same time I rolled violently back to the right and had just begun a hard turn in that direction when I saw another MANPADS launch low and off my nose. I continued my break turn and kicked out more flares, grunting under the G-forces as I twisted around in the cockpit and strained to keep my eyes on both missiles. I was in a real pickle. You need to watch a missile to best defeat it, but the different directions that the two missiles were coming from, and which way I was turning, meant that I couldn't keep my eyes on both. I had the feeling that the first missile had already begun to lose track of my jet, maybe following the flares I had dropped, so I decided to concentrate on the second one. I guess that was a good call, because even though I don't know where that first missile went, my jet didn't blow up! As I watched in the second missile, I quickly realized it wasn't guiding on me either, and it passed harmless by my left wing and up into the night sky. The AAA was now arcing all over the sky near my jet. The SA-6, which probably wasn't here anyway, hadn't decided to play, but I was sure that there would be a lot more MANPADS launches if I kept doing this. I was also out of gas, so I decided that it was time to call it quits for the night. I quickly turned off my external lights, turned away from the city, and climbed away into the night sky. Without the visual references to locate my airplane, the AAA trailed away and soon stopped almost completely. I also didn't see any more MANPADS launches. Maybe the Yugoslavians thought they got me? Or maybe they too could sense that the night's fun was over. Who knows?

The whole flight back to Aviano I ruminated over the mission. Sure, it was an exciting mission, and the Wild Weasel pilot in me enjoyed the tactical problems it presented, but I was disappointed that we didn't find anything worth blowing up. I have never minded hanging my ass out if the end result was worth it, but to go through all of that just because the French were scared didn't set well with me. I guess in the end it must have had some value, because the French resumed flying around Pristina after that. Maybe I had embarrassed them as more than I embarrassed the Yugoslavians?

It was after midnight by the time we landed back at Aviano. The crew van picked me up from my jet, and then picked up Bobcat 83 next. As the crew van trundled along to pick up the rest of our flight, Bobcat 83 sat down on the bench seat next to me and leaned his head back against wall of the van. He blew out a long and loud breath before saying, "Flaring low and slow over Pristina with your lights on. Gutsiest move I ever saw, Maverick." I had to smile...after all, who doesn't like a good Top Gun movie quote!

Chapter 8: A pick-up game

The morning started out pretty normal. We were staying in a place up in the Dolomite Mountains, in a town named Piancavala, at a hotel that normally catered to skiers and tourists. Now it was filled with pilots from our squadron since there wasn't enough room at Aviano Airbase to billet everyone during Operation Allied Force. I woke up in my hotel room just before my alarm clock was scheduled to go off. Before showering and shaving, I took a few minutes to just look out my window at the snow-covered slopes and mountains…it was beautiful! I always knew a war would be a good one based on my accommodations. During Desert Storm I lived in a villa near the Red Sea and went snorkeling every day after flying. The Army and Marine guys like to make fun of how easy we Air Force pilots have it, always staying in nice hotels, in great locations, while they live in tents and foxholes. It also always made them jealous that we raked in the per-diem too. It was good-natured ribbing, but at the end of the day nobody drafted them into the Army or Marines, they volunteered for that life!

After a shower and shave, I pulled on my flight suit, grabbed my flight jacket, and walked downstairs to meet up with the three guys who would be in my flight that day. We grabbed cappuccinos from the hotel bar and then drove a rental car down the mountain to Aviano to go fly. The drive down the mountain was every bit as dangerous as the combat missions we flew. Snowy mountain roads in Italy are narrow and full of hairpin turns and edged with no-guardrail cliffs. But the real danger came from whoever was driving. We had this awesome bet going in the squadron. We kept track of the quickest times both up and down the mountain. As I've mentioned, fighter pilots are a competitive bunch, so whoever was driving saw death as a small price to pay for a shot at the immortal glory in winning the title of fastest driver on the mountain! I was never as scared in the jet while being shot at as I was on trips up and down the mountain as a passenger. It was a stupid thing to do, risking our lives like that. We were lucky that nobody skidded off a cliff. We did have one car run straight into the side of the mountain one day...more about that later, but fortunately no one was hurt. For the record, when the war was over and we headed back to South Carolina, I held the fastest time up the mountain!

The guys I was flying with that day were Walnut, Melvin, and Web. Walnut and Web were outstanding fighter pilots, and I was happy they were going to be my wingmen that day. Melvin, well, not so much. Melvin was a great guy, but I had already learned that he wasn't cut out to be a combat fighter pilot. When you fly combat, most of the guys perform just like they do in training, which is to say they do a solid job putting warheads on foreheads. But there are a few that surprise you, in good and bad ways. There are some guys who are average in training but become real killers when the time comes. And there are a few who, in training are outstanding, but underperform badly when the shooting starts. Melvin was one of those who over-promised and under-delivered. Again, he was a great guy, and I would love to have a beer with him tonight and catch up on old times, but he never made the shift from peacetime thinking to combat. Let me give you an example.

A few days before, I was leading a 4-ship in the KEZ, and Melvin was my #3. The weather was dogshit, with huge, towering thunderstorms all over the place. We were supporting some A-10s, who were working some bad-guys over pretty good, and to cover them I set us up in the best weather I could find, which was in a lane of clear air between two big walls of active cumulonimbus clouds. These thunderstorms were spitting lightning and drowning the ground beneath them with deluges of hard rain. As we were covering the A-10s, Melvin called up on the radio and started quoting a regulation that said we weren't supposed this fly close to thunderstorms. Knowing the regulation better than he did, I replied that it said no "peacetime" mission required flying near thunderstorms, and that the bad-guys getting shredded by the A-10s right now would definitely agree with me that things weren't peaceful at the moment. Melvin continued to whine, and I eventually called him out. I told him over the radio, for everyone to hear, that if he was too scared of the weather, he was welcome to fence out and wait for us on the tanker. That shut him up for the time being, but after the mission he still insisted he was right. He was adamant that we should have made the A-10s go home so we could all avoid the bad weather. Unbelievable. As it turned out, Melvin was just as painful to deal with on the mission to come.

The date was 21 April 1999, I was flying tail number 91–395 on mission number 3431 foxtrot alpha, and my call sign was Glock 41. Originally, we were scheduled to fly on a KEZ Mission. The flight consisted of myself and my wingman, Walnut. Glock 43 was Melvin and Glock 44 was Web. When we got our weather brief before going out to the jets, it showed Kosovo as being socked-in with low clouds and rain. I honestly wasn't surprised that before we could even taxi out to the runway, our mission was canceled due to the bad weather. Everyone shutdown their jets, and we piled back into the crew van and drove back to ops. We had no sooner turned in our flight gear when the squadron operations desk called us. We were being re-rolled to another mission, and we needed to go over to the 555th Fighter Squadron, the "Triple Nickle," ASAP to plan it.

We drove over to the Triple Nickel operations building, and after asking around, we found the guys we would be flying with. Our flight consisted of 2 F-15E Strike Eagles, callsign Cuda 91, and a single EA-6B, callsign Spear 1. I was delighted to learn that we would be flying the first daylight raid into downtown Belgrade, and our target was a bridge. Normally one of the guys who would be dropping the bombs would be designated as the Mission Commander, but none of the F-15E crews were qualified, so that title went to me. The Mission Commander is like the flight lead for the whole mission. He decides on the final plan and is responsible for quarterbacking the whole thing in flight. Cool!

We didn't have much time to organize the mission, and standing around a map, we quickly came up with a game plan we all agreed was tactically sound. Normally, when we flew to targets in northern Yugoslavia, we either came in from the west across Croatia, or came in from Hungary to the north. But the tankers we would use were the ones originally tasked for our mission in the KEZ to the south, and for some reason they couldn't reposition from their refueling tracks over Albania…which was way down south. This was no big deal for the F-15Es and the EA-6B, because they can carry an ass-load of gas. But for us it was a big deal, as F-16s don't carry a lot of fuel. Tanking over Albania, then flying south to north across all of Yugoslavia, fighting, and flying all the way back to Albania wouldn't leave us much extra. I would have to watch our gas very closely, especially if we got into a real fight, which was very likely over Belgrade in broad daylight.

The basic plan we agreed upon was for us to come off the tanker in a long trail formation. EA-6Bs are SLOW airplanes, so I would push Spear 1 out well ahead of the rest of us, and we would overtake and pass him enroute to the target area. By the time we reached the target area, he would be in the perfect position and distance behind us to jam the enemy's radars. My 4-ship would split up, with me and Walnut out front, the Strike Eagles behind us, and Melvin and Web behind them. As we approach the target area, I would take my two-ship and breakaway to the east. Cuda flight would flow in and strike the bridge in Belgrade with AGM-130 air-to-ground missiles. Melvin and Web would remain behind Cuda flight to protect them from any SAMs or AAA that might pop up near them. My two-ship would arc around to the east, checking out Obrava Airfield for MiGs and SAMs, and then turn north to sanitize behind the F-15Es and Melvin and Web as they came off target and headed south. This was important because Belgrade was home to a couple of MiG bases, and I didn't want any MiGs to take off and sneak up on everyone as we exited the target area. I would take my 2-ship north searching for MiGs, and then turn around to follow the trail of jets back to the tanker. Because the EA-6B was so slow, he would wrap up his stand-off jamming and leave ahead of everyone else. That way we could protect him too, catching up to him about the time we reached Albanian airspace. It sounds like a complicated plan, but it was really quite simple. Everyone was happy with it except Melvin, who whined that we needed to talk about it more. His complaining made the F-15E guys roll their eyes at each other.

We were in quite a rush because the tankers couldn't wait for us forever. And of course, when we got back out to our jets, everything seemed to go wrong. Three of our four F-16s developed maintenance issues right after engine start, and the pilots had to run to spare aircraft. My jet seemed fine, but as soon as we taxied my INS dumped. Between this, and the first mission I flew in Operation Allied Force, you might be thinking that INS failures were common in the F-16, but that's not true. They are actually kind of rare. I guess I was just lucky! Rather than abort the jet, I first tried to re-align the INS while I was taxing out. The whole time I was thinking, "Why do the INS's seem to hate me?" Fortunately, the INS worked just fine after being realigned. Eventually all of my flight joined me at the end of the runway, and we took off about 15 minutes later than planned.

The follies continued, however. Once airborne we found out that our tankers had cancelled, but we were directed to continue on our way and the AWACS would find us the gas we needed from other tankers in the airspace over Albania. In the end it all worked out, and about 45 minutes after launching we found ourselves saddling up behind a KC-135E. Because we were 15 minutes late, I elected to refuel me and Walnut first so that we could push off and lead the airplanes north in trail of Spear. Melvin and Web would get their gas last and catch up to the train of aircraft. We had no sooner left the tanker, when the AWACS, callsign Magic, called everyone back and told us to remain on the tanker until further notice.

Things like this happened all the time, and you rarely found out why. Years later I did time at the Combined Air Operations Center in Qatar, where we ran the air wars over Iraq, Afghanistan, and the Horn of Africa. It was there that I watched the Generals and Colonels do the same thing to aircraft on combat sorties over and over again. Rather than trust the pilots to have built a good plan, the Generals and Colonels would have the fighters hold on a tanker for hours at a time while they dithered, worried that something might go wrong, and they wouldn't get their next promotion. But during this particular Operation Allied Force mission, I was ignorant of dithering Generals and Colonels, and was blissfully content to fly around with the tanker and sight-see. After about an hour Magic passed along that we were cleared to proceed. While everyone topped off their gas, I did my Mission Commander thing and set a new time over target for the strike package. Everyone pushed off the tanker, and we were on our way. The plan was finally in motion!

The flight north across Yugoslavia was pretty uneventful. Looking to the east toward Kosovo I couldn't see any of the bad weather that scrubbed our original mission. Our formation eventually overtook the EA-6B, and before long, we were approaching Belgrade. Spear 1 broke off and began orbiting, ready to jam any radars that came up in the target area. At the designated spot I turned my 2-ship to the east. The weather was beautiful, and everything was going as planned as I began to search in earnest for MiGs and SAMs. Everything seemed quiet, and before long I heard Cuda flight launch their missiles at the bridge. At just that moment Walnut and I were approaching Obrava Airbase. I was staring at the runway, looking for any MiGs taking off, when I saw a donut-shaped ring of dust blow up off the ground...a SAM launch...an SA-3 by the look of it. I had no indications in my airplane that I was targeted, and I quickly ascertained that Walnut was not targeted either. I lost track of the SAM as it flew off into the sky like an errant Roman Candle. Following the plan, Cuda called that they were off target and beginning a turn to the south. That was my cue to turn my 2-ship to the north and keep an eye behind Cuda and Glock 43 and 44. I knew that Cuda would fly close to the SA-3 site, so I commanded Glock 44, Web, to shoot a HARM at the airfield where the SAM was located. It was really a magnificent site! Seconds after Web called "Magnum", his HARM streaked up into the stratosphere, leaving a sharp, fast-moving contrail. The HARM arced up and eventually got high enough it exited the contrail layer, only to begin leaving another contrail as it tipped over and back into the contrail layer on its dive to the target. The contrail from the HARM passed above me, and I jerked my attention away from staring at it when Cuda called that they had SAM indications behind them, back towards Belgrade. I picked up the speed of my crosscheck, searching for the SAM with my equipment, looking for MiGs with my radar, and for MiGs and SAMs both with my eyeballs in the sky around us. Sadly, I didn't see anything except for Cuda and Glock 43 and 44 as they flew underneath me southbound. With a final look for MiGs, I swung my 2-ship around to the south and rejoined line-abreast with Glock 43 and 44.

It was at this point that I realized Spear 1, the EA-6B, hadn't left before us as planned. This meant that instead of him being out in front where we could keep an eye on him, he would quickly fall behind us. I thought about this for a second, because EA-6Bs are basically defenseless against MiGs. They are defenseless not only because they are slow and can't turn for shit, but also because they don't carry any air-to-air ordnance. My first reaction was to simply orbit my 4-ship for a while and let Spear 1 get a good distance ahead so we could then follow him out. But a quick survey of the gas situation ruled that out. Our F-16s would barely have enough gas to make it back to the tanker with a small reserve. So, there wasn't much I could do; Spear would have follow us out. Not optimum, but I thought to myself, "Hey, MiGs are so rare there's no chance Spear is in any real danger. He'll be fine." So south we flew, and about half-way across Yugoslavia we were a good 15 miles in front of Spear. And that was when the AWACS, Magic, came over the radio with a warning that set my pulse to racing.

"Glock 41, Magic, probable hostile, bullseye 200 for 37, Angels unknown." It took me about 2 seconds to process that. The AWACs was saying there was an airplane out there that was most likely a bad guy, and the location Magic gave was about 20 miles behind Spear at an unknown altitude! Even though we were critically low on fuel, there really wasn't any question about what I needed to do. Our job was to protect the strike package from MiGs and SAMs alike. We didn't have the gas for it, but Spear 1 would be raw meat on the table for any MiG. "Glock 41 flight, hook right," I called over our inter-flight radio. As we wheeled around, I figured we would find and kill the MiG, and then make for the runway at the U.S. Army occupied airfield in Tirana, Albania. Hopefully we could make it there before we flamed out. If not, at least we stood a good chance of bailing out over a somewhat friendly nation when the tanks ran dry. On the strike frequency I called out, "Magic, Glock 41 flight commit, commit...picture north." Magic came back with the exact same call he issued before, and I concentrated my radar search in that area. I could see Spear 1 on my radar, but nothing else. Knowing that most of the time you get killed by someone you don't see, my head was on a swivel. I was also on the lookout for a "SAM-bush," with the MiG being used as bait to draw us into a SAM's lethal envelope for a quick shot. To improve our tactical situation, I commanded Glock 43 and 44 to maneuver away to the west a little bit and climb, and I turned my 2-ship to the east some and descended. This would build lateral and vertical separation between our airplanes and force the MiG to choose a side. When he did, the targeted airplanes could turn away, and the other F-16s could turn behind the MiG for a quick kill.

As we continued northbound, I still wasn't seeing anything on my radar scope. I knew my wingmen weren't seeing anything either, or they would have said something. Looking outside, I watched Spear 1 pass down my left side, roughly co-altitude, going in the opposite direction. I was now in a position to keep myself between Spear 1 and any bad guys, and that made me feel a lot better. I decided to get an update from the AWACS. "Magic, Glock 41, picture." After a few seconds the radio crackled, "Magic, picture clear." This didn't mean there were definitely no MiGs, only that he couldn't see them. Now I was really worried. Had the MiG gotten by us at low altitude where the AWACS might have also missed him? Was he so close to Spear or one of us that the AWACS couldn't tell the difference between us and him? In a hot second, I directed Glock 43 and 44 to spin back and follow Spear. I continued north for another minute or so, banging away with my radar, and straining my eyes in a vain search for that MiG. I eventually decided to turn around and follow everyone back out. A part of me wasn't surprised that I couldn't find anything, since the AWACS might have seen something anomalous on his radar, and not an actual airplane. But I also had enough experience with the NATO AWACS to know they were not as good as the all-U.S.

crews, and they could very well have simply lost track of the MiG. As we flowed south to the tanker track, bringing up the rear, my neck began to ache from constantly checking six!

As if the potential for an undetected MiG wasn't enough to worry about, we were almost out of gas. I checked how much fuel each member of my flight had and decided we would just be able to make the tankers over Albania. As we fenced out and switched our radios over to the refueling frequency, I got another surprise. Magic informed me that our tanker had run out of gas and returned to base! "Un-fucking-believable!", I said to myself. "Are the guys in the AWACS *trying* to kill us?" Getting mad wasn't going to help things, so I keyed the radio and asked Magic if there was another tanker with some extra gas. Before the AWACS could answer, a voice responded, "Glock 41 this is Tuna 41, our fighters cancelled, so we can help you out."

After refueling we cruised back to Aviano AB. The sun was in the west and reflecting off the sea as we cruised home up the east side of the Adriatic. Before long the Dolomite Mountains came into view, and we began our descent to land. Once we were on the ground, we debriefed with intelligence, and decided to go to the Officer's Club to celebrate our part in the first daylight attack on downtown

Belgrade. While we were sipping our beers and waiting for our steaks to arrive, I asked everyone what they thought of the mission, and if they had any questions. To my amazement, Melvin decided to chime in with another of his "peacetime" rules that we broke. This time he was unhappy that we had turned around and gone after the MiG with no idea of its altitude. Under peacetime training rules, everyone has assigned blocks of altitude that they remain within in unless they have awareness of where the adversary is. This rule is important, as it minimizes the chances of a midair collision if nobody sees each other. But in combat it is entirely possible, and maybe even probable, that you won't know the exact altitude of who you are fighting. So, you need to be careful. I sipped my beer in disgust. "So, Melvin, you're saying we shouldn't have turned around to save Spear, just because we didn't know what altitude the MiG was at?" He started to defend his point and I cut him off. "If you think that way you need to stop flying fighters and go teach at pilot training." Melvin stared at me, clearly angry at the put-down, but he stayed quiet. The rest of dinner he didn't say much, even when the mood lightened around the table. After that, I made sure Melvin and I never flew in the same formation again.

So, like I wrote earlier, some guys were never able to make that mental transition from training to combat. I'm going to leave this story with a word of advice to any future fighter pilots out there. Its advice that I used to give to my young pilots when I was the Operations Officer of the 55th Fighter Squadron later in my career. In peacetime there are rules of engagement that are designed to keep pilots safe while training. Every one of those training rules is written in the blood of guys who died because of how dangerous training for combat can be. When you go to war, the training rules don't apply anymore. That said, when you do something in combat that you wouldn't do in training, you need to recognize it, and keep in mind why it's a peacetime no-no. A famous fighter pilot once eloquently wrote, "The way to long life as a fighter pilot is to fly with your brain and not your muscles." Another less famous pilot I used to fly with said it less eloquently, "Don't be all balls, dick, and   no   forehead."

Chapter 9: Air-to-Air

The daylight raid into Belgrade, and the resulting engagement with the imaginary MiG, made me realize that I should probably take some time to talk about air-to-air combat. Every fighter pilot wants the chance to shoot down an enemy airplane. There's no question that dropping bombs and shooting missiles at ground targets is a lot of fun, but at the end of the day, every swinging-dick who flies fighters wants to be an "ace." To become an ace you need to shoot down five airplanes in aerial combat. WWI, WWII, and the Korean War generated a lot of aces. In Vietnam there were only a small handful. And since Vietnam, a U.S. fighter pilot getting the opportunity to shoot down an enemy airplane in combat has become a pretty rare thing. In the last 40+ years there have been no U.S. aces, and only two U.S. pilots have shot down three enemy airplanes. One was a guy flying an F-16, and another flying the F-15. Operation Desert Storm was the last time the U.S. shot down any significant number of jets. Incidentally, I was airborne on the opening night of Desert Storm, and the first MiG kills of the war happened just to my north, but that was the only air-to-air action I was close to in that conflict. In reality, the U.S. military is a victim of its own success in so many ways. The U.S. military produces the finest and most capable fighter pilots in the world, and we savaged the Iraqi Air Force so badly, that no country since has been dumb enough to challenge us for air superiority. Rather than lose all of their expensive fighter planes trying, our adversaries simply hide their airplanes as best they can and wait out the war. But every now and then someone gets froggy and tries to challenge us. Being the lucky guy who gets challenged is every fighter pilot's dream! I never got to shoot down anyone, but I came close a few times. Before I regale the reader with tales of my almost glory in aerial combat, it is worth taking the time to describe how the U. S. Air Force trains fighter pilots for air-to-air combat.

The journey to becoming a would-be ace starts in pilot training. Back when I started pilot training, we took our first flights in the T-41 Mescalero (the Air Force version of the Cessna 172) in a program call the Flight Screening Program. Note that the title of that course includes the word "Screening" and not "Training." The Flight Screening Program is all about determining if you have the aptitude to fly, and not about really training you to be a pilot. I attended the Flight Screening Program at the U.S. Air Force Academy. On the day I reported for training I already had about 1,500 hours of flight time, and held my private pilot and commercial pilot licenses, with multi-engine and instrument ratings. I was pretty confident I could solo in a Cessna 172 in the 10 flights allowed. Indeed, I had no problems with the course, and I actually enjoyed the program. But as advertised, not everyone in my class made it through the screening process; about 25% of the class washed out due to a lack of aptitude.

After completing the Flight Screening Program, I received orders to Vance AFB, in Oklahoma, to attend Undergraduate Pilot Training, more commonly referred to as UPT. At UPT you join a class of other student pilots from all walks of life, including some from other countries. My class had two students from the Kuwaiti Air Force, and three from the Spanish Air Force. UPT takes a full year to complete. At the end of that year, you get your wings and an assignment to fly a certain airplane. On paper, at least, every Air Force UPT graduate is universally assignable, and can go fly any airplane in the inventory. In reality, from day one, you are taught that "cream rises" and what you get to fly after pilot training is purely dependent on how well you perform compared to your classmates. While the numbers vary slightly from class to class, only about the top 15% will rank high enough among their peers to choose a fighter if that's what they want. So, if you have ever heard that fighter pilots have competitive personalities, you now know the path to becoming a fighter pilot hinges on being the best and beating those around you. It pays to be a winner.

UPT starts out with a heavy academic load, covering all of the topics a military pilot needs to know. These include classes on aerodynamics, navigation, weather, aerospace physiology, how jet engines work, etc. You also take a lot of classes that teach you about the systems in the airplane you will fly next. For most of the year you have class for a half a day, and either fly the jet or a simulator the other half of the day. At the end of every academic course there is a test, and your scores contribute to your class standing. Every flight and simulator mission receives an overall grade, with individual scores for dozens of sub-areas. Every single one of those grades goes into your class standing as well. Most days on the flight line begin with a written pop-quiz on a random topic that you should know.

Not only does that pop-quiz grade contribute to your class standing, if you flunk the quiz, you don't fly that day. Absolutely every day begins with an evolution called Stand-Up. The instructors would select a student, and call out, "Stand Up Lieutenant Snuffy!" You never knew in advance who they would call, but everyone was called on equally over the year at UPT. The Lieutenant in question would quickly proceed to a table in the middle of the room, carrying only their checklist and inflight guide (the things you would have in the jet), and stand at attention. Next the instructors would turn on the overhead projector, and then read a short script, describing what you are hypothetically doing in the jet, the weather, and other details. An example might be, "Lieutenant, you are in your T-37 on a solo ride out in the working area. The weather is 7,00 broken with 9 miles visibility back at Vance, and the winds are from 270 degrees at 12 knots. You are performing aerobatics, and at the top of a loop a red light comes on in the cockpit. This is what you see." And then they would turn on the overhead projector and put up a display of your instrument panel showing all of the dials and switches. They would then say, "Lieutenant, you have the jet!" You would reply, "Roger, Sir, I have the jet!" And then you were expected to assess the information available to you and describe in great detail what you would do in response. This might involve reciting certain emergency procedures from memory, and then looking at the instruments to determine what was wrong, and then working through the correct checklists. The scenario would end at a "logical" conclusion, which might be landing the jet, or even ejecting. If you were doing well, and the scenario was a long one, you might have the instructor interrupt you and say, "Lieutenant Snuffy, have a seat." You would then let out a sigh of relief and set down, while the instructor would call on another student to "Stand Up" and continue where you left things. If, however, things were going badly, the instructor would stop you with a "Lieutenant Snuffy, sit down." "Sit down" was a bad thing to hear, and meant that you had busted Stand-Up, and wouldn't fly that day. And of course, your performance at Stand-Up mathematically contributed to your class standing, just like everything else you did at UPT.

As if all this wasn't enough pressure, you were never more than three days from washing out of pilot training. If you failed something, you would get a second chance the next day. If you flunked that redux, you would go to an elimination ride. If you failed at that, you were removed from training and sent to a review board to determine your future in the Air Force. You might think that all of this pressure and intentional stress was unfair, but in reality, it enabled the standard of excellence expected of Air Force pilots. The theory is that if you can't handle the artificial stress of answering questions or flying a jet in a training environment, you won't perform well when you are really in a jet with an engine fire, or when someone is shooting at you. And it's true. Thinking clearly at one G and zero knots while standing in the middle of a classroom might be stressful, but it's nowhere close to the pressure of doing the right thing when your life is at stake! At the end of the day, anyone who graduates Air Force pilot training has proven that they can think clearly under stress.

At UPT the first airplane I flew was the T-37B. It was affectionately nicknamed "The Tweet" due to the insanely high-pitched sound of its ancient turbo-jet engines. The Tweet was an outstanding primary flight trainer. The cockpit was arranged with side-by-side seating, with the student on the left, and the instructor on the right. Each pilot had a full set of instruments and controls. Despite the cluttered cockpit and having two underpowered jet engines, the Tweet was a delightful airplane to fly, and its flight characteristics were very forgiving. The first phase of T-37s was called the Contact phase. In Contact, you learned how to take off and land, and to also perform aerobatics. This included putting the jet into spins, which is something I have heard they don't teach Air Force pilots anymore (which is a shame). A lot of people are scared of spins when they first do them, but after a few they actually become a lot of fun. Intentionally spinning most modern military airplanes is a prohibited maneuver, but if you do find yourself in one it sure is nice to have had some practical experience getting out of it. But I digress. Other T-37 training phases included Instrument Flying, Low Level Navigation, Cross-Country Navigation, and Formation Flying. Every Phase contained a very specific set of skills to master in a fixed number of flights. Every Phase also ended in a check ride, and as with everything else at UPT, your scores on each check ride contributed to your class ranking.

After completing the T-37 half of UPT, my class moved on to fly the T-38 Talon. The T-38 may very well be the damn sexiest airplane ever built. There is an old saying that T-38s look like they are going fast even when they are parked. It's true! The T-38 is s supersonic jet, with razor-thin wings and two engines that come equipped with afterburners. It's the first jet you get to fly that requires you to wear a G-suit, and I must confess that I felt like a real stud the first time I zipped on my "speed-jeans" and climbed into the Talon. In a T-38 the student sits in the front cockpit, and the instructor flies from the backseat. The jet was introduced as an advanced trainer back in the 1960's and was designed to prepare students for follow-on assignments in the Century Series fighter, like the F-100, F-101, F-102, F-104, F-105, and the F-106. The airplane was truly high-performance. It could go supersonic quite easily, had a roll-rate of 720 degrees per second with full aileron deflection, and was very unforgiving in the traffic pattern. The T-38 portion of UPT was broken into the same Phases as T-37 curriculum...Contact, Instrument, etc. It was a real hotrod, and if you learned to fly it safely, you probably could fly any airplane in the inventory. If you don't know what a T-38 looks like, put this book down and go look it up on the Internet.

So, what does all of this do with why Air Force fighter pilots are good at air-to-air combat? Well, the price of entry into the fighter community is being exceptional at all of the skills taught in UPT. To graduate at the top of your UPT class and get a fighter, you have to be good at academics. You have to be able to think and perform under pressure. And you have to be able to fly the hell out of an airplane, be it flying in formation, doing aerobatics, or in the clouds at night. Having good "hands" is assumed to be the price of admission for would-be fighter pilots.

After UPT my next stop was a course called Lead-In Fighter Training, or LIFT. I attended LIFT at Columbus AFB, in Mississippi. At LIFT we flew the AT-38, which was a T-38 with a cool paint scheme and modified with a basic gunsight, as well as having hardpoints to carry practice bombs and a gun pod. LIFT is a short class, lasting only a couple of months. The syllabus is intense, and it gives you your first taste of what fighter pilots really do, and what is expected of a young wingman. The first rides in the program were to get you current again in the airplane. Most LIFT students will have been several months without flying between their last flight at UPT and arriving at LIFT. In between, you attend ejection seat water survival school, land survival school and prisoner of war training, and a few other courses, including a trip to the centrifuge.

Everyone hates going to the centrifuge. Modern fighter planes can pull very high Gs for extended periods of time. Passing out under high G forces is almost always lethal. So, before you can show up to fly an F-16 or an F-15, for example, you have to be "certified" in the centrifuge to be able to stay awake under high G. The centrifuge is exactly what it sounds like. It consists of a generic airplane cockpit at the end of a mechanically driven arm that spins around really fast, placing a precise amount and duration of G forces on the poor schmuck strapped inside. While G forces are technically the same no matter how you experience them, the reaching of those Gs in the centrifuge is nothing like being in a real airplane. The spinning motion of the centrifuge causes a great deal of spatial disorientation, and while most pilots won't puke in it, they do climb out dizzy and perhaps nauseated. The effect is so pronounced and common, that pilots are not allowed to fly a real airplane for something like two or three days afterwards because of lingering vertigo concerns.

The Air Force has a couple of centrifuges, and I went to the one at Brooks AFB in Texas. This particular centrifuge was the oldest one, and due to its primitive construction, was also considered the most miserable one of all. Now, historically, tall skinny guys have a harder time pulling Gs than short stocky guys. When I went to the centrifuge, I was six feet tall and about 130 pounds. When it was my turn to get ready for my ride in the centrifuge, a crusty old senior enlisted guy pulled me aside into a room alone. He looked me up and down and said, "Sir, you're not going to make it through the centrifuge...you are way too tall and skinny." Up to that point I hadn't been worried because I had never had any issues pulling Gs before. Without saying another word, he reached in a desk drawer and pulled out a roll of two-inch-wide strapping tape…the kind that has threads of nylon running in it to make it almost unbreakable. He squatted down in front of me and proceeded to wrap bands of that tape around my legs, both above and below each knee. He wrapped it as tight as he could pull it. No shit, it was like having two tourniquets put on and each leg, and I immediately went numb below the knees. He stood up, nodded, and said, "That should help you keep the blood in your brain. Now put on your G-suit and don't tell anyone." "I nodded and added a "thank you" as I donned my G-suit. With my lower legs completely numb I could barely walk, and the this guy kept a hand on my arm as he walked me to the centrifuge. As we approached the centrifuge cockpit another pilot from my UPT class was getting out. He was as white as a ghost, sweating, and shaking like a leaf! The guy who got him out of the cockpit led him by, and I will never forget the look in my friend's eyes…they were wide open like a cartoon character being electrocuted! I was scared to death as I got strapped in.

The door to the centrifuge shut, and I assessed my situation. The "cockpit" had no windows, but at least it was vaguely configured like an F-16. The seat was reclined 30 degrees, and there was a side stick. In front of me was a horizontal bar of colored lights, along with a camera facing me. There was a speaker somewhere in the cockpit, and the centrifuge controllers began reading a script explaining what would happen. Basically, I would "fly" a series of centrifuge runs, each of a pre-determined duration with a pre-programmed rate of G onset, as well as a preset G-limit. When I was ready, I could initiate each run by pulling on the side stick. As long as I held the stick full aft the profile would continue until completion, with me straining against the Gs to stay awake. The horizontal bar of different colored lights was there to help me assess how I was doing. As the Gs increased beyond your ability to deal with them, you would begin to lose color vision, then you get tunnel vision, and eventually will black out. The idea was that if you have an effective G-straining maneuver you won't do more than maybe loose color vision while looking at the horizontal light bar. If you started to no longer see the lights at the outer ends of the bar, it was a clue you were getting tunnel vision, which indicated the Gs were getting the better of you. If you couldn't see the lights at all, it meant you were taking a G-induced nap!

G forces cause a pilot to pass out because they cause the blood to drain from your brain and pool lower in your body. To stay awake under high G, a pilot must tighten the muscles in his legs, butt, and stomach, and push against a deep breath, to pressure blood up into your brain. To be honest, it is pretty much the same thing you do when you have to push really hard to force out a shit. Sadly, the long-term result of G-straining all the time is it promotes hemorrhoids, a problem common to fighter pilots. We also learned in our centrifuge academics that under 9-Gs, your heart and lungs sag four inches down into your body. I'm pretty sure stretching all of your internal plumbing like that isn't good for you either. Interestingly, the weight of your blood pooling in your lower body can create so much blood pressure, that it also causes the capillaries in your skin to pop. When this happens, you are left with little red dots all over your ass and legs. They look like measles, and since they are G-induced, they are known as "Geasles". The packing tape tourniquets that the SNCO had tied around my legs would help me stay awake under high G because the blood in my body couldn't sink all the way to my feet. My G-strain would only have to push the blood up from my

171

knees instead of all the way from my feet. Of course, I would be wearing a G-suit too. A G-suit fits very snugly around your legs and stomach, and it inflates like a blood-pressure cuff does on your arm when the airplane pulls Gs. When the G-suit squeezes you, it helps push blood out of your legs and gives you something to push against to make your G-strain more effective. A G-suit adds about two Gs to your tolerance.

The first profile was a warmup consisting of a slow onset of constantly increasing Gs. During this first centrifuge run you weren't supposed to strain at all so that you can determine your "resting G-tolerance," which is defined as the point you can't see half of the lights on the bar. I seem to recall that they told us the average person has a resting G-tolerance of around four to five Gs. When I pulled on the stick for the first time, I immediately understood why everyone hated the centrifuge. As is spun up to the rotational speed needed for the G required, I immediately felt like I was tumbling head over heels and rolling around sideways all at the same time! With no visual horizon to help your brain and inner ear reconcile things, the vertigo was intense. I don't remember for sure, but I think my resting G-tolerance was around five Gs or so. Once that profile was completed, the voice over the loudspeaker told me to let go of the stick, which brought the centrifuge to a halt very quickly. That sudden cessation of motion induced another, different, and much worse vertigo sensation. I'm pretty sure I said something out loud, like, "Fuck me!" Instantly I wasn't worried about passing put under G...I was hoping I wouldn't puke all over myself!

For about the next 10 minutes I worked through a set of profiles with ever increasing G loads and durations under G. Arguably, these profiles simulated the on again, off again, nature of Gs while dogfighting. I didn't have any issues pulling the Gs, but by the time I got to the final profile I was glad it was almost over. The last profile was the hardest one, with a high G-onset rate, reaching 9Gs for something like 30 seconds while looking over your shoulder and talking "pilot talk." I have never been so thankful for anything as when the centrifuge came to a stop for the last time, and the voice over the speaker cheerfully said, "Good job, we'll come get you out!"

I sat there in the darkened cockpit, gasping for breath, and trying not to puke. The door opened and the same guy who put me in helped me climb out. The vertigo was still bad, and I guess he expected that, because as I stood up out of the centrifuge, he pushed me up against the wall and said, "Hang on the wall, Sir, while I get the cockpit set for the next guy." He then helped me back to the same office as before, where he cut the strapping tape tourniquets off my legs. The pain of the blood flowing back into my legs was a wonderful sensation, as it took my mind off puking from the vertigo. In my long career that followed I spent a lot of time at 9-Gs, and it never felt like that centrifuge!

My centrifuge story has a couple of humorous addendums. First, after my centrifuge ride was debriefed and I was told I had passed, I called Base Transportation for a ride back to my quarters. As I waited for the ride, I sipped a little water, and stood looking out a window to help my inner ear recover from the insane vertigo I was still experiencing. "The Air Force is right to not let you go fly a real airplane after this," I thought to myself. Before long the base taxi arrived, and I very slowly climbed in with my G-suit slung over one shoulder. I told the driver which building they had billeted me in, and off he drove. The motion of the car was instantly too much, and I frantically rolled down the window and leaned out, puking up the water I had sipped earlier. The driver didn't even slow down. "Don't worry, Sir, we see this all the time." I gave him a weak thumbs-up and continued to hang half-way out the window for the whole ride, dry heaving every time the car slowed, sped up, or made a turn. After what seemed like an eternity we arrived at my quarters. I mumbled a thanks and began walking to my room. The door to my room was open, and the maid was changing my bed linens. As I got to the door another wave of vertigo washed over me, and I literally lost my balance and fell face-first across the threshold of the doorway. I was so dizzy I couldn't stand up, so I began crawling on all fours, dragging my G-suit behind me. The maid was walking toward me, and I thought that she was getting ready to freak out over a pilot collapsing and crawling into the room. I was getting ready to explain to her what had happened, when she stopped and looked down at me very seriously. She took a deep breath and laughed out loud! She then stepped across me and walked out the door. I rolled over and looked back at her, kind of surprised by her lack of empathy. As she reached back to pull the door shut, she smiled and said, "We see this all the time!" I laid there on the floor for a few minutes and decided that I would crawl into the bathroom and just lay on the cool tile for a while. It was everything I hoped it would be!

The second humorous addendum to my centrifuge story happened several years later. At the time I was the Chief of Scheduling for the 78th Fighter Squadron. The squadron's scheduling shop builds the weekly flying schedule, and to do that also collects and posts all of the training and meetings people need to attend as well. One of the other pilots in the squadron (a really great guy, awesome fighter pilot, and now a Lieutenant General well on his way to a fourth star) was a favorite target of my practical jokes. Only a few weeks earlier I had stuffed little pieces of banana into the fingers of his flight gloves one day before he went to fly. The look on his face was hilarious when he pulled the gloves on right before engine start! So, one day I was thinking about how much the centrifuge had sucked, and that I was thankful it was a "one and done" event. You only have to repeat the centrifuge if you leave flying for more than three years. It dawned on me that every pilot thinks the centrifuge sucks, and it would be a great joke to make someone believe they were going to have to go back again. After thinking it through, I decided to play this prank on the guy I mentioned earlier. I knew that he, like every pilot in the squadron, would come in on Thursday afternoon to see what he was scheduled for during the upcoming week. So, on Thursday, after the Operations Officer had officially approved the schedule, I erased this guy's name from all the flights he was scheduled to fly and marked him as away on temporary duty (TDY) for the week. My bait was set, and all I had to do was wait for him to take it. Sure enough, he eventually strolled in, and after staring at the schedule for a while, he came over to my desk and asked, "Hey Axe, what's this TDY thing I'm scheduled for next week?" Without looking up from my computer, I said, "Uh, that's your centrifuge recurrency trip. I think you are going to Brooks AFB for it." It was all I could do to not laugh as he slowly turned back to the scheduling board and just stood there. "What do you mean, recurrency training?" Pausing a moment to steady myself from laughing, I casually replied, "Yeah, well, I heard they just started this recurrency thing. Anyway, your name popped up for it." After a minute or so he walked out of the scheduling office without another word, but from the look on his face I knew he was stressing out about it! After he left, I had a good laugh, and then erased the TDY from the board and put him back into the flights for which he had been previously scheduled. About 15 minutes later he came back in. "Hey Axe. I just talked to Admin, and they don't have any travel orders for me to the centrifuge." I stopped what I was doing on the computer and looked up at him, giving my best distracted look, making it obvious I had been concentrating on something else, and only half heard him. "Huh? Oh, yeah, it got cancelled. I will let you know when you get rescheduled." With that, I went back to my work, and watched in amusement as he walked off. I smiled, pleased in the knowledge that he never knew it was a joke, and that over the coming weeks and months was sure to check the upcoming schedule with trepidation at finding his name next to another centrifuge recurrency TDY! To this day I have never told a soul about that joke, and I hope my old friend reads this book one day as finds out he was pranked.

And now I will return to describing LIFT. As I mentioned before digressing about the centrifuge, the first couple of flights at LIFT are simply about regaining your landing and instrument approach currency in the jet. You also got to do a lot of aerobatics and advanced handling maneuvers to regain your feel for flying the jet near the edges of its performance envelope. Once you were back in the saddle, so to speak, it was time to start learning how to use an airplane as a weapon.

The first half of the course was dedicated to introducing Basic Fighter Maneuvers, or BFM. BFM is more commonly called "dogfighting," which is when two fighter planes fight against each other, one on one. BFM is truly the sport of kings, and I guarantee you there is nothing two fighter pilots enjoy more than facing off against each other to prove who is the better pilot. It's like the Thunderdome from that Mad Max movie… "two men enter, one man leaves." But unlike in the old days when fighter pilots just had to learn through getting their asses kicked by more experienced pilots, we now teach BFM in a very structured way. Believe it or not, there is a lot of academic learning when you first approach BFM. You learn how to assess engineering diagrams and analysis of how your airplane performs. You learn what it does well, and what it doesn't do well. And you also study the same information about other types of airplanes, both allied and enemy. If you can't grasp what the engineering tells you, and then apply it in the air, you won't be a fighter pilot for very long.

In the air, learning to dogfight starts with what we call Offensive BFM. In Offensive BFM you start out behind your opponent at pre-determined distances, speeds, angles, and altitudes. When both pilots are ready, a "Fight's On" call is made over the radio and the dogfight begins. In Offensive BFM, from that position of advantage, your goal is to maintain the offensive, solve problems of aspect, angle, range, and closure, and kill your opponent with a valid guns track. Along the way you take all of the missile shots you can, but you can only call a "kill" with a valid gun shot. To be good at Offensive BFM you have to fly aggressively and be able to place your jet at an exact point in time and space to make a gun shot.

The follow-on phase to Offensive BFM is Defensive BFM. As the name suggest, in Defensive BFM you start out with your opponent behind you at a predetermined set of parameters. At the "Fight's On" call, your job is simply to survive. You do this by defeating the initial threat, which is most likely a missile shot of some type. Even though your opponent won't call you dead unless they kill you with the gun, you must train yourself to recognize and react to the different missile envelopes in addition to the threat of the gun. After defeating the immediate threat, you maneuver your jet so as to create problems of aspect, angle, and closure that make it hard, or even impossible, for your opponent to maneuver into position and score a guns kill. If you are really on your game and create enough problems, you might even be able to outmaneuver your opponent, and place yourself in a position where you can seize the initiative and go on the offensive yourself. Regardless, you goal is to survive as long as possible.

The final phase of BFM training is High-Aspect BFM. "Aspect" is the angle between the longitudinal axis of your two airplanes. As an example, two airplanes heading in the same direction have low aspect. And if two airplanes are pointed nose-to-nose at each other, they are high-aspect. So High-Aspect BFM begins with the fighters pretty-much nose-to-nose. Nominally this means that neither airplane has an inherent positional advantage. This means you have to execute a game plan that maximizes your strengths and exploits your opponent's weaknesses and mistakes, with the goal of reaching a position where you can transition to flying Offensive BFM.

There are some important things to consider about BFM, in all of its flavors. First, a pilot's experience almost always means more than how advanced of a jet they are flying. The World War I ace, the Baron von Richthofen, famously wrote, "It is the man, not the crate, that matters." The U.S. military has the most lethal fighter pilots in the world because we train harder and more realistically than anyone else. And it shows whenever we train BFM (and everything else for that matter) against the pilots of other nations. Technology can narrow that quality gap, but it hasn't eliminated it quite yet. Another factor is what kind of fighter plane are you flying BFM against. Most of the time we trained against other airplanes of the same type, but we also regularly had the chance to fly BFM against different airplanes too. This is called Dissimilar BFM and is a vitally important part of training a good fighter pilot.

As noted, before, fighter pilots spend a lot of time in the books studying the performance and weapons capabilities of other fighters. When you fly Dissimilar BFM, you have to be able to make the fight happen in ways that take advantage of your jet's strengths and exploits your opponent's limitations. A great example comes from World War II. Japanese Zeros were fantastically maneuverable, and they could turn tighter and climb faster than the American P-40s. At first, American pilots were getting their asses handed to them because they would let the Zeros draw them into turning engagements. But soon they learned that their P-40s could dive to much higher speeds than the Zero, and with that resulting speed advantage, they could zoom their airplanes back to higher altitudes faster than the Zeros could climb. The American pilots in their P-40s changed tactics, and began making diving attacks against the Zeros, refusing to turn with them, and instead diving away and zooming up for another diving attack. By playing their strengths against the enemy plane's weaknesses, the American pilots won more fights than they lost. So, whenever I found myself in a dogfight against another jet, I needed to tailor my tactics accordingly. I would fight an F/A-18 very differently than I would fight an F-15E. And if I had ever found myself high-aspect with a MiG-29 or a J-20, well, I knew how I would force them to fight my game as well!

The second half of LIFT was dedicated to dropping bombs. The AT-38 had very simple gunsight that didn't include any sort of computerized weapons delivery functions. Basically, you were dropping bombs the same way they did from fighter planes in World War II. You would be surprised how hard it can be at first to even drop a bomb within 100 yards of the target!

Before I move on to how we learned air-to-air in the F-16, I will tell a flying story from LIFT. All airplanes, even fighters, have limits on how many Gs they can pull without bending or breaking the airplane. Guys have actually pulled so many Gs that they have literally ripped the wings off of airplanes before. Fighters have three G limits you must abide by: positive Gs, rolling Gs, and negative Gs. Rolling Gs occur when you roll the airplane while also adding positive or negative Gs at the same time. Rolling G limits are usually much lower than pure positive Gs limits. One day at LIFT, I was flying an Offensive BFM ride. The ride was going great, and as I saddled up for my last fight, I noticed that I wouldn't have enough gas for the whole fight. Being a savvy student fighter pilot, I pointed this out to my instructor in the back seat. He asked what my plan was, and I told him that just before my fuel state reached the point where I needed to go home, I would execute a maneuver called a "High-Angle Guns Separation," or HAGS. From my offensive position behind the bandit, I would pull my nose way in front of my opponent's flight path in a very aggressive bid to shoot him with the gun from high aspect. A smart opponent would recognize my aggressive move and turn hard into me to complicate my gun shot. In doing this I would sacrifice being offensive and pass the bandit almost head-on. With the bandit heading in the opposite direction, I could dive away from the fight before he could turn back around and gun me. So, at the "Fight's On" call I slammed the throttles into full afterburner and pulled my nose way out in front of the bandit who was turning into me. As expected, he recognized my bid for a high-aspect gun shot and turned even tighter into me, trading his airspeed for a smaller turn radius. At the perfect moment, I slammed the stick back the other way and pulled it back at the same time, to roll into my dive and separate from the fight. Now, the rolling G limit in the AT-38 is something like 4 Gs. The positive G limit is just over 7 Gs. I executed my HAGS so violently that I pulled 9 rolling Gs! The instructor was pissed. After we landed the jet had to undergo a major inspection to determine how badly I had bent it. This inspection required the jet to be almost completed stripped to the frame and x-rayed for cracks as well. Our entire LIFT class went out to see it, and we posed for a group photo in front of the stripped-down jet sitting up on jacks in one of the maintenance hangars. On the day we graduated we gave the LIFT squadron a blown-up copy of that picture, with the matting around it signed by each of us. We also added a little plaque at the bottom of the picture that read, "Over-G'd and Under-Achieved...the story of our lives!"

So, with LIFT completed my next assignment was to Luke AFB, Arizona, to learn to fly the F-16. Luke AFB was home to the active-duty Air Force's F-16 Replacement Training Unit (RTU) that included five dedicated RTU squadrons, and a squadron that provided follow-on training in targeting pods, night vision goggles, and Forward Air Controller-Airborne qualifications. Luke also hosted the U.S.-administered F-16 RTU for the Singaporean and Taiwanese air forces. As if Luke wasn't busy enough with those eight training squadrons, it was also the home to a squadron of Air Force Reserve F-16s as well. For a new pilot coming to the F-16 from UPT the course was eight months long. The syllabus started with a heavy academic load to learn the F-16's systems, which were considerably more complex than those in the T-37 or T-38. The first three or four rides were in a two-seat F-16 and focused on learning how to take off and land the F-16, perform aerobatics, and gain instrument proficiency. After soloing, I soon found myself taking an instrument check ride, which qualified me to fly the F-16 while learning to use it as a weapon. The remaining flights in the program were broken into two basic phases: air-to-air and air-to-ground. The last few rides of the program brought together everything we had learned, and the whole class flew as wingmen for the instructors while conducting large force exercises that simulate real strike packages. When I went through the RTU the last ride was a check ride, granting you "Combat Mission Ready" status as a real, honest to God F-16 wingman. In later years the mission ready check ride was moved out of the RTU and to your gaining unit, reflecting the need to be proficient in the specialized primary missions of most F-16 squadrons.

Air-to-air training in the F-16 started off with the same sequence of flights we experienced at LIFT: Offensive BFM, Defensive BFM, and High-Aspect BFM. From there we moved on to more complex tactics, beginning with Tactical Intercepts, or "TI." In TI we started by learning to run single-ship radar intercepts against a single bandit.

Piggy backing on that, we expanded into being a wingman in a flight of two F-16s running radar intercepts against a single bandit. Mixed in with the TI lessons we flew Air Combat Maneuvering missions, referred to as "ACM." In ACM we would work as a two-ship against a single bandit, learning how to work as a team to efficiently kill and survive. Think of ACM as two against one dogfighting. Eventually we took our air-to-air training to even higher levels, conducting two against two and four against four missions called Air Combat Training, or ACT. Later, after learning how to drop bombs effectively, you combined all of your training in simulated missions. The F-16 is a multi-role fighter, meaning it can do more than one thing on the same mission. We would fight other airplanes to get to the target, drop our bombs, and then fight our way home. Even after I was a combat mission-ready fighter pilot in a front-line squadron, I still trained in all of the basic air-to-air building blocks. So hopefully you now understand why the U.S. military, and the Air Force in particular, produces the most highly trained, proficient, and lethal fighter pilots in the whole world. And it also sets the stage for me to talk about my real-world air-to-air experiences!

My first really close call to shooting down an enemy fighter came while flying an Operation Southern Watch mission. Southern Watch was the name of the no-fly zone enforcement over southern Iraq. In the summer of 1996, I was in the 78th Fighter Squadron, and we were the Air Force's on-call fighter squadron in the U.S. for rapid deployment to the Middle East. We were all carrying around beepers, and had a bag packed so that we could deploy on a moment's notice. The idea was that we could move the squadron anywhere in the world and be flying combat missions within 72 hours. One of our sister squadrons at Shaw AFB, the 79th Fighter Squadron (the "Tigers"), was already in the Middle East on a scheduled Operation Southern Watch rotation, flying out of Dhahran Air Base in Saudi Arabia. On the 25th of June 1996, terrorists detonated a truck bomb there in an attack known as the Khobar Towers bombing. Tragically, 19 U.S. service members were killed and 498 injured, including many from the Tigers. The 79th was declared combat ineffective, so our beepers went off and, in a few days, the 78th was in Saudi Arabia picking up the Wild Weasel mission. Everyone knew the Iranians were behind the Khobar Towers bombing, so tensions were pretty high.

One of the first missions I flew was what we came to call "The Dawn Patrol." Like the name implies, we would launch out of Prince Sultan Airbase in Southern Saudi

Arabia so that we could be on station over Basra, Iraq, as the sun came up. While not an official tasking, whoever flew the Dawn Patrol would sonic-boom the city of Basra every morning at official sunrise to let the Iraqis know we owned the sky over their country. The painful part of these missions was getting up around 1am to make a 3am takeoff so that we could refuel on a tanker and be on station in time. Combined with the jetlag of having flown half-way around the world a few days earlier, I was really tired when we took off that morning. On this mission we were flying with the callsign Spalding 11 and 12. My flight lead that day was tired too, so he told me to not bother rejoining into a close formation after takeoff. Instead, he instructed me to just hang back a couple of miles and follow him to the tanker. After climbing up to our cruise altitude I set the autopilot up and sat there looking out at the starry night sky.

The next thing I knew, I woke up with a panicked nap-jerk...I had drifted off to sleep! The shot of adrenaline that went through my body as I realized I had just woken up from a nap in an F-16 brought me instantly awake and alert. My first thought was, "Oh God, how long was I out, and where am I?" If I had been asleep for more than about 45 minutes, I would have been halfway across Iraq and perhaps almost out of gas! The first thing I did was look at my gas totalizer, and I breathed a big sigh of relief to see that I had plenty, which means I had probably only drifted off for a few minutes at most. My next concern was where my flight lead was located. My last memory was that he was slightly off the right of my nose and a couple of miles away. I searched that area, but he wasn't to be seen. I knew he couldn't be too far away, so I started searching the sky around me, and sure enough, I found him off the left side of my jet, around the 10 O'clock position. Now fully awake and relieved that all was well, I slipped back into trail formation behind my flight lead without saying anything. Soon we reached the tanker, got gas, and sonic-boomed Basra at sunrise.

That morning we didn't have any fighters to escort for the first half-hour or so, and to kill the time we set up a combat air patrol (CAP) just to the northwest of Basra. After a few minutes, the AWACS let us know that two Iranian F-14s had been scrambled in response to our presence and were being vectored toward us. We re-oriented our CAP into an east-west orbit and began searching for the Iranian jets. It wasn't long before we had them on our radar scopes. The ROE dictated that we couldn't fly into Iranian airspace, but if the F-14s locked us with their air-to-air radars, even across an international border, that would constitute a hostile act, which would mean we could shoot them down in self-defense. As we flowed east toward the F-14s, they turned and pointed straight at us.

Everything that unfolded was exactly like how we trained for air-to-air. This was basically a two against two TI setup, and I was hopeful it would lead to me and my flight lead blasting both Iranian jets out of the sky! My flight lead called on the radio, "Spalding 1 targeted, Spalding 2 sort." This let the AWACS know he had the F-14s identified, and he was telling me to use my radar to target one of the F-14s in accordance with a pre-briefed plan that ensured we didn't both lock onto the same bad guy and let the other one get away. Using the radar controls on my throttle I slid the cursor over the two radar contacts, and then centered it over the one I would take responsibility for. With the press of a button on the stick I commanded my radar to lock onto the target. "Spalding 2 sorted" I replied. At this point I reached over and turned my Master Arm Switch to "Arm" so that my missiles and gun were live.

Glancing back at the radar I could see that the F-14s were about 40 miles away and still pointed straight at us doing 500 knots, which was about the same speed we were flying. With 1,000 knots combined closure we were less than two minutes from merging with them, and probably just seconds away from their firing missiles at us from beyond visual range (BVR) if that was their plan. My thumb was hovering over the "pickle button" on the stick to fire my missiles at the first indication of the Iranian jets locking me with their radars. Suddenly, my radar showed that the F-14s were beginning to maneuver. I watched as they executed a 180 degree turn and ran away from us at high speed. With the Iranian border approaching, Spalding 11 handed responsibility for tracking the F-14s back to the AWACS, and we turned around ourselves. We stayed in our east-west CAP for a while, but the Iranian F-14s flew back to their base and landed without any further action. I guess they were content to play until we locked them up, and they chickened out...which was disappointing. If they had taken things just a little further, we would have killed them both.

My second close call with a shootdown came in 1998 during another Operation Southern Watch deployment, this one to Bahrain. On this particular mission I was leading a four-ship, callsign Zipgun 51. Along with four F-15Cs, callsign Cylon 21, we were going to support a U-2 as it flew over Baghdad. These missions were usually more work than they were fun. The U-2, while it flies very high, also flies very slow compared to F-15s and F-16s. So, to stay in a good position to watch the U-2 as well as the different SAM sites and MiG bases, both us and the F-15s had to fly a sawtooth pattern that required a lot of cooperation and timing for everyone to be in the right position. At any given time two F-15s and two F-16s would be following the U-2 around, while the remaining F-15s and F-16s went back to the tanker for gas. When they returned, we would swap out following the U-2s so that we could go refuel. This would go on for a couple of hours.

On this particular day we were right over the southern suburbs of Baghdad, when the AWACSs, callsign Darkstar, called out that a MiG-25 "Foxbat" was taking off out of Al-Taqaddum Airbase, which is about 75 miles west of Baghdad. The MiG-25 is designed to fly at high speeds and high altitudes, and its primary mission is to shoot down high value airborne assets (HVAAs...pronounced "Have-uhs") like the U-2, the AWACS, and the RC-135. Over the years since the end of Operation Desert Storm, the Iraqis had tried on several occasions to run a MiG-25 south to catch an AWACS or an RC-135, but they had never gone after a U-2 before. Most of the time the HVAA would run away, and the MiG-25 would also run away as U.S. fighters moved to intercept it. Once, an F-16 actually shot one down. But this occasion was a lot more serious, because there was no way the U-2 could outrun a MiG-25 at this short of a range. Upon hearing that this Foxbat was airborne, the U-2 immediately turned south, and my two-ship, along with Cylon 21 and 22 snapped into action.

In the past when a Foxbat launched, it would fly a predictable course, initially climbing to the north and then turning and accelerating back south as it made its run for its target. And today, as Darkstar gave a running commentary on the MiG, it was obvious they were following the same ground track. As soon as the U-2 began bugging out, Cylon 21 and 22 turned south as well, with my two-ship about 15 miles in trail. Technically, the Foxbat hadn't done anything that would allow us to shoot him down yet. He needed to commit a hostile act first. There were a lot of ways he could commit a hostile act, but the most likely would be locking his radar onto someone, shooting at someone, or crossing south of the 33rd latitude line and into the no-fly zone. With the U-2 running south, the first opportunity the Foxbat would have to commit a hostile act would be crossing the 33rd parallel as he chased the U-2. By heading south behind the U-2, the F-15s were building enough separation from the Foxbat to be able to turn around and shoot the MiG as soon as it crossed into the no-fly zone. My two-ship was saddled-up 15 miles behind the F-15s so that we could not only back them up against the MiG, but to also attack any SAMs or AAA that popped up in the event this was an ambush using the MiG as bait.

Soon, Darkstar called out that the Foxbat was turning south. A few seconds later Cylon 21 and 22 turned back north, having mentally determined that the timing was right to execute their plan. As we turned north behind Cylon, I quickly found the MiG on my radar, and monitored him as I searched for SAMs and AAA. Darkstar was on the radio providing updates to the Foxbat's position, altitude and heading, when I began to realize that Cylon had messed up the timing of their turn back to the north and were going to have to turn around to avoid flying past the MiG before it would cross the 33rd parallel. In a heartbeat I realized the glory of the kill would be mine, as I was some 15 miles behind Cylon and would be in a perfect spot to shoot as the Foxbat crossed the 33rd parallel! I could hear the frustration in Cylon 21's voice as he called "out left" and passed the MiG off to me. "Zipgun 51 targeted," I replied, and I shoved the throttle into afterburner. Air-to-air missiles like all the speed they can get before launch, and my plan was to launch mine supersonic. I reached over and armed my missiles, and decided that when I could shoot, I would dust the rails…firing all four of my AMRAAMs and Sidewinders. I needed to do this since Foxbats are incredibly fast, and if I missed with my first missile, he would blow by me and be out of range before I could turn around and shoot again. Besides, as rare as the chance is to shoot down a MiG, I wasn't going to worry about conserving my missiles!

Now I was climbing at Mach 1.1, straight at the MiG, who was just seconds from crossing the 33rd parallel. I had the capability to monitor this precisely and was touching my thumb on the pickle button to shoot when the AWACS shouted over the radio in a near panic, "Darkstar declares FRIENDLY, FRIENDLY! United Nations nuclear inspector flight to Kuwait!" My thumb froze. Still supersonic I zoomed toward the now friendly aircraft, and in seconds I could make out the distinct shape of a Boeing 737 off my nose! As I blasted past the airliner I came out of afterburner and pirouetted my jet so that I rolled out above and behind it. I slowly overtook it until I was flying along-side it. The pilots were oblivious, and never noticed me sitting off their wing. All I could think was "Holy crap! I almost blasted an airliner out of the sky!" I followed the

737 for a while, and eventually the AWACS sent the U-2 back north and we continued the mission. In the days to come the near tragedy was dissected. We figured out that the Iraqis found out we were aware of the ground track that Foxbats would fly. In a sneaky attempt to trick us into shooting down a U.N. weapons inspector flight, the Iraqis has vectored the 737 on the common MiG route of flight as it left Al-Taqaddum, and simultaneously made bogus radio calls like they were vectoring a MiG after the U-2. It almost worked. No kidding, I believe that if Darkstar had delayed their "friendly" call by only a second or two I would have hammered down on that pickle button and launched my weapons. I've often wondered if the folks on the 737 ever knew how close they came to dying. I guess it's better to be lucky than good!

My first air-to-air encounter of Operation Allied Force came one day while flying in the KEZ. I don't seem to recall flying many early in the day missions during Operation Allied Force. Maybe because it was a NATO air war, and the Europeans generally don't like getting up early. But on this occasion, I was over Kosovo in the vicinity of Pristina around 0600, supporting a flight of A-10s looking for targets of opportunity. My callsign that day was Blade 21, and Blade 22 was a very young wingman, only a few months removed from his initial training in the F-16. Everything was quiet when we arrived on station, so I set up our CAP so that we could watch Pristina Airport and cover the A-10s at the same time. It was going to be a beautiful clear day, but the sun was still low enough in the sky that the rolling terrain and forested hills cast long shadows across the ground. I was keeping a close eye on my young wingman and writing myself a note on my kneeboard about his formation keeping when the AWACS called. "Blade 21, Magic, intermittent hits low along the road between Leskovak and Nis. Possible helo." "Blade 21 copies," I replied. On our inter-flight frequency, I turned our formation back to the east and reminded Blade 22 to check that his weapons were armed, and his video tapes were turned on. Then I leafed through the different pages information I had strapped to my leg until I found my map; I knew where Nis was located, but Leskovak was a new one for me. It turned out that Leskovak was a little town just south of Nis, and the road between them was somewhat nestled down beneath some low hills that ran along each side. After studying the map for a few seconds, I came to the conclusion that in all likelihood, the AWACS had not seen a helicopter, but rather a car driving fast down the road. Sometimes, when viewed at the right (or wrong) angle a car going really fast can bounce back a doppler return that is misinterpreted by a radar as a low-flying airplane or helo. Still, it was worth checking out, especially since nothing else was going on.

Even though I was confident this was a non-event, I still proceeded with caution. Nis was known to have active SAM batteries, as well as an airfield that could harbor MiGs. To get my best look with the radar I brought our flight across Leskovak from the south. This would allow me to look down the length of the road without the hills on each side blocking my radar's view of low-flying objects. It would also give my equipment the best chance of detecting any SAM radars. The Yugoslavians had a lot of road-mobile SAM systems, and it wouldn't be surprising if they had one along this major north-south road in the hopes of catch a NATO jet by surprise. I had no sooner focused my radar down the road when I too began to get hits on a low-flying aircraft of some sort! I keyed the mic and called out, "Magic, Blade 21 contact low, my nose, 15 miles. Blade 21 commit, commit!" This let the AWACS and everyone else on that frequency know I had found something and was maneuvering to intercept it. The raw returns on my radar scope were consistent, so I went ahead and locked the contact up. My radar had no problem establishing a lock, and as it began tracking the target, I could look through my Head's-Up Display (HUD) and see that what I had targeted was far enough off to the side from the road that it was most certainly not a car!

I was actually close enough to the target now that I needed to bunt the nose over so that I could see the area where it was flying. Diving down from 25,000 feet, I was having a hard time seeing any detail on the ground, mostly due to the shadows created by the low sun angle. I leaned my head to the side to look around the HUD, hoping that I without the reflective glass of the HUD in the way I could see better. And as soon as I did, I saw it. There was a small collection of white colored buildings, and I happened to be looking at them exactly when the helo flew over them. He was low, skimming the rooftops, in a left-hand turn. He looked like he was slowing down too, so I suspected he was getting ready to land. "No worries, "I thought to myself, "I can strafe him if needed." As the helo flew away from the white buildings, he almost became invisible against the trees and shadows. I knew that if I took my eyes off of him, I would lose sight of him and might never find him again. Fighter pilots call this being "padlocked." Keeping sight was important, because in accordance with the ROE, I would need to visually identify

(VID) him to declare him hostile and kill him. I shoved the nose over harder, planning to drop down to tree-top level. I planned to blast by him, get a good look, and then pirouette up and around to shoot him. I needed to let my wingman know what was up, so I called out, "Blade 21 padlocked my nose five miles, in for the VID." My wingman responded immediately with, "Blade 22 SA-6 bearing 360. Blade 22 is blind."

This meant that my wingman had indications that an SA-6 SAM was somewhere in front of us, and his "blind" call let me know that he didn't know where I was...he had lost sight of me. My guess was that he was busy looking at the SA-6 indication when I dove for the ground, and he was still flying up at 25,000 looking for me. A wingman's number one responsibility beyond not crashing his own airplane is to keep visual with his flight lead. Barreling downhill, 45 degrees nose-low at 500 knots, passing 10,000 feet, I had to make a decision fast. If I looked away from the helo to find Blade 22 with my own eyeballs, I would lose track of the helicopter and might never see him again. There was also the possibility that the helo was bait to draw us in for the SA-6 to shoot at us, a tactic called a "SAM-bush." My first thought was, "Fuck him, if he is blind right now, he is on his own." But my wingman was a very new and inexperienced guy. If I left him alone where he was, he might very well get blasted out of the sky by that SA-6. I just couldn't justify the risk. I yanked my F-16 into a high-G climbing right-hand turn, kicking out a few flares in case there was a guy down there with a shoulder-fired SAM looking to shoot me. Over the radio I called, "Blade 21 flight hook right, reference south. Blade 22 say altitude." My wingman replied, "Blade 22, 25,000." Squinting up into the clear blue sky I quickly caught sight of Blade 22, high above and turning to the south as I had directed. Climbing away I checked six for any SAMs and called the AWACS. "Magic, Blade 21 is out south, target the helo if able." Magic acknowledged but replied that he didn't have any radar awareness of the helo anymore. While we flew south, I climbed up to 25,000 and talked Blade 22's eyes back onto me. Once we were back in a good tactical formation, I turned us back up the road to look for either the SA-6 or the helo. We searched until we were out of gas and needed to go back to the tanker, but we never saw the SAM or the helicopter again.

My second air-to-air encounter of Operation Allied Force also began in the KEZ. It was a night flight, and my callsign was Luger 51. Luger 52 was one of my best buds, and a highly experienced fighter pilot. As an aside, it is always nice to fly with another highly experienced guy. You don't have to worry about them as much as you do with the younger pilots, and if things go to hell on a bobsled, you know they can take care of themselves and do the smart thing. Anyhow, we were orbiting just to the southwest of Pristina on a beautiful starry night. We were supporting a pair of A-10s, Bazooka 11 and 12, who were searching for targets of opportunity but having no luck. Everything was quiet, and there was never a hit on the radar scope or a sniff of an enemy radar anywhere. We had just turned west in our orbit, and I looked down at my radar scope and saw something odd. Bazooka 11 and 12 were right where I expected to see them, about 25 miles off my nose in an orbit of their own at 15,000 feet. But about five miles beyond Bazooka there was another radar hit, moving north at 20,000 feet. We weren't supposed to have any other fighters show up, and I hadn't heard anything on the radio to make me expect otherwise. I quickly keyed the mic asking the AWACS if he saw anything, "Magic, Luger 51, do you show any other flights on station besides me and Bazooka?" The NATO AWACS controller replied in a thick Turkish accent, "Magic, negative." Right away my "Spidey-sense" began to tingle!

In a combat zone locking someone up with your fire-control radar is a big deal. Everyone is flying around with live air-to-air missiles and locking someone is basically the same as pointing a loaded gun at them, so you don't do it lightly. But all of the information I had told me this new contact wasn't a NATO airplane, so I slid my radar cursors over him and locked him up. The contact was at 20,000 feet, going 420 knots, tracking north. I keyed the mic once more, "Magic, Luger 51 picture, single group seven miles west of Bazooka at 20,000, track north, declare." Asking for Magic to "declare" the contact meant I needed the AWACS to determine of it was friendly or hostile. Under the ROE I was not able to declare the contact hostile unless I could VID him as an enemy aircraft, which would be impossible at night since we didn't have targeting pods or night vision goggles. The AWACS, however, had the ability to make that call on their own. If I could get a hostile declaration and get in range, I could shoot him down. The AWACS controller replied, "Magic, picture clear." I assumed the controller didn't know where I was looking, so I provided another, more detailed description of exactly where the contact was located. Again, the AWACS controller said he saw nothing there. At this point I knew I needed to take matters into my own hands. Keying the mic, I called out, "Magic, Luger 51 commit, commit. Bazooka, bugout south." Bazooka acknowledged, but the AWACS came back once again with a "picture clear" call.

The range and the angles to the unknown contact were less than optimum for me to intercept him. Thinking it through, I was sure it must be a fighter, given the speed and the fact that the AWACS couldn't see him. I also knew there were no NATO planes in central or northern Yugoslavia that night, so it was probably a bad guy going somewhere, since he flew right by the A-10s without stopping to engage them. Drawing a mental line north from the contact it dawned on me that Obrava Airfield was in that direction. Perhaps that was where he was going in such a hurry. This made sense to me, since we had heard many intelligence reports of the Yugoslavians taking airplanes off at night from airfields near their borders and running them back to safer location like Obrava. Without a second thought, I turned right and pointed at Obrava in the distance, and slammed my throttle into afterburner, looking to cut off the contact's escape route. Luger 52 was saddled up three miles behind me, and I made a quick double-check that my weapons were armed and my video tapes recording. As we raced toward Obrava I tried again and again to get a declaration from the NATO AWACS, but he continued to profess that he saw no other airplanes on his radar scope. Very quickly I had indeed cut off the contact, and popping the throttle out of afterburner, I rolled out a mere two miles behind it while matching its speed. I was in perfect position to shoot this guy...if only I could get a hostile declaration! "Magic, declare bogey my nose two miles!" "Magic, picture clear." I pounded my fist against the canopy in frustration! The contact began descending straight toward Obrava. I had no indication of any SAMS or AAA, but I was very cautious about getting dragged into an SAM-bush. I finally had to accept that I wasn't going to get a hostile declaration out of the NATO AWACS, so I turned our flight aside to stay out of range of any SAMs at Obrava. As we arced around the airfield, I saw them turn on the runway lights for a brief time, and then the airport went dark again. Luger 52 and I headed back south to the tanker for gas before going home for the night. I kept checking six and asking the AWACS for a picture to our north, in the hopes that the Yugoslavians might try and send MiGs after us, but all was quiet. After landing at Aviano AB, we included the unknown contact and our chase in our after-action report to the intelligence folks. A few days later they let me know that it was indeed a MiG-21 I had chased. Frustrated that I couldn't get a declaration of any sort, I called the NATO AWACS squadron and talked to one of the U.S. Air Force AWACS controllers assigned to the unit. After hearing my story, he empathized and promised to look into it, although I never heard back from him. He did share an interesting tidbit of information, however. He said that many of the NATO guys in the squadron disagreed with the air campaign against Yugoslavia, especially America's interference in an internal European matter. In his personal opinion he thought it was entirely possible that the Turkish controller saw the MiG but was simply not going to cooperate with me.

My last air-to-air encounter of Operation Allied Force happened the day after one of my squadron-mates actually did shoot down a MiG-29. His flight of four F-16s was leaving the Belgrade area after covering a strike package when an unknown aircraft was detected in the area they had just been working. The weather was pretty bad, and after getting unhelpful, and even confusing information from the NATO AWACS controller (sound familiar?), he was able to identify the target as a MiG-29 and shot it with a pair of AIM-120s at fairly short range as he was diving out of the clouds. That night we had a huge squadron party at the hotel in Piancavallo. The owner of the hotel,

"Rocky," celebrated with us. He enjoyed himself so much he decided to break out his jug of homemade Grappa. I had never heard of Grappa before, and while Rocky went to fetch the jug one of the pilots in the squadron, who was married to an Italian lady, explained that this was basically moonshine. He also added that if we declined drinking his Grappa, Rocky would be extremely insulted. I wasn't too worried since I have imbibed my share of moonshine over the years. Even if Rocky's batch wasn't very good, one shot wouldn't be too much to bear in the name of politeness to our host. A short time later Rocky returned with a huge glass jug under each arm. He proudly explained that he had flavored this batch with orange peels, and it had been buried behind the hotel for several months to let it age and smooth out. Rocky then rolled out a serving table on wheels that was loaded with full-size water glasses and proceeded to pour each of us an entire glass full of his orange-flavored Grappa! All around the room the guys were looking at each other nervously. The thought of drinking a full water glass of moonshine was pretty intimidating, especially since it had been made clear to us that not drinking it would be an insult. Rocky led with a toast to our squadron's MiG-29 kill, and I hesitated to see how the other pilots would react when they took a drink.

The first few who took a sip visibly gasped and choked after swallowing it. Rocky noticed too, laughed, and raised his glass with a hearty "Salute!" before gulping down his whole glass. In a display of incredible bravery, all of the pilots began drinking their glasses down too. I took a deep breath, placed the glass to my lips, and gulped it empty as fast as I could swallow it. As the last mouthful was going down, I almost died. Rocky's Grappa had to be nearly pure grain alcohol, and it burned like fire. Add to that a very strong orange peel flavor, and it was all I could do not to heave it all back up. The room grew quiet except for the sound of Rocky laughing at us. My eyes were watering but looking around the room I was glad I drank the whole thing at once. Several of the guys were trying to get it down with a break between mouthfuls, and they were having a hard time convincing themselves to keep going. Eventually everyone emptied their glasses, making Rocky happy, and maintaining the squadron's honor. The party kept going, but pretty soon the alcohol I had just guzzled really hit me, and I thanked Rocky for his hospitality before staggering off to my room. I needed to sleep this off since I was flying tomorrow.

The next morning, I felt surprisingly good. In fact, I didn't have a hangover at all. I wasn't flying until later that afternoon, so I ate a big breakfast and spent the morning doing laundry and writing a few letters home. Around 11:00am I threw on my flight suit and met my wingman in the hotel lobby to go in for the day's mission. I drove us down the mountain and soon we were at the base, where we grabbed lunch and briefed our mission. Our callsign that day was Razor 31, and we were going to provide Force Protection to a strike package hitting targets just west of Belgrade. The weather was looking pretty bad, but the mission proceeded anyway. The strike package, consisting of my two-ship and four F-15Es, took off together and proceeded over the Adriatic Sea, and then turned east across Croatia and into Hungary where we met our tankers to refuel. Leaving the tanker we followed the F-15Es into Yugoslavian airspace. The weather was a solid undercast from about 10,000 feet down. The F-15Es needed to be able to see their targets since they were dropping laser guided bombs, and based on the unbroken clouds below, they cancelled their mission and headed back to Aviano. My wingman and I, however, were tasked to fly around over Yugoslavia to "provide presence." I was okay with this, since without a strike package to protect we could roam around a bit and see if we could find any SAMs or early warning radars to attack.

My plan was to work the area along the western edges of the cities of Novi Sad and Belgrade. We made a few orbits, but the Yugoslavian weren't cooperating, and we didn't find any radars to go after. I decided to swing us around and cruise along the southern side of Belgrade and see if we could find something there. This would take us into the same area where our pilot shot down the MiG-29 the day before. It crossed my mind that maybe we would have some luck there too.

No sooner had we turned eastbound just south of Belgrade, when I saw some radar returns down low, tracking westbound about 30 miles away. Whoever they were, they were down beneath the weather hugging the ground. I was hopeful that today I would be able to get the AWACS to give me a hostile declaration, so I keyed the mic and contacted the AWACS. "Magic, Razor 31, declare single group, 10 miles south of Batchanikya Airfield, low." After a few moments, the AWACS controller replied with "Magic shows picture clean." With less than 30 miles between us and the target, and the distance closing rapidly, I either needed AWACS to make a declaration or I was going to have to dive down through the weather and VID it. I asked the AWACS for another declaration, giving him a very detailed description of where the unknown airplane was located, but he still couldn't (or wouldn't) see it. I wasn't too keen on diving through the thick weather to low altitude to conduct a VID. In the clouds I would not be able to visually acquire SAMs or AAA shooting at me, nor would my wingman and I be able to provide visual support to one another. At that moment my radar began to break out two contacts flying in a two- or three-mile trail formation. That made up my mind for me, I was going after them.

I locked my radar onto both aircraft and targeted my wingman to the trailer. Looking at what my radar was telling me, there was no doubt these were airplanes. They were skimming the ground at about 230 knots, so there was no way they were cars on a road, and at that speed they probably weren't helicopters either. I rolled my jet upside down and pulled the nose to the ground, double checking my missiles were armed. Because we would be coming through the weather my wingman would be behind me, and I was concerned that he might shoot a missile past me. I didn't want him shooting a missile past me in case the missile got confused and came after me instead, so I said something to him on the radio as a caution. My plan was to pop out of the weather just a few miles in front of the bogies and go high-aspect with the lead aircraft to get a good VID. As soon as I could positively VID the lead airplane as hostile, I would shoot his wingman in trail. I would then turn around behind the lead airplane and shoot him too.

As we dove down, I keyed the mic again to let the AWACS know what I was doing. "Magic, Razor 31 and 32 are engaged my nose, 10 miles, two-ship in trail, in for the VID." "This is it," I thought to myself, "I am finally going to get and air-to-air kill!" Instead of a simple "Magic copies," which is what I expected, the AWACS controller jumped in frantically. "Razor 31 Magic declares that group friendly, friendly...it's two C-130s." If I didn't have an oxygen mask on my jaw would have dropped into my lap. What the fuck were two defenseless C-130s doing flying around in broad daylight just south of Belgrade? And if they were indeed C-130s, how come the AWACS controller didn't know about them when I initially reported them as unknowns? I pulled my F-16 nose up and zoomed back to 25,000 feet with my wingman close behind. At this point I wasn't really sure what to do, so I led us around the area for the next 20 minutes, watching the C-130s on my radar as they droned west and out of Yugoslavian airspace toward the Adriatic coast. Eventually our time on station was up and we returned to Aviano. The only thing I could think of was that the C-130s were part of a Special Forces operation. Since we weren't originally scheduled to be south of Belgrade, that could explain why we weren't informed of tier presence. That would also explain why the AWACS controller played dumb, claiming he didn't see anything until we were about to attack them. Regardless, it was a real let-down to once again be denied an air-to-air kill.

Chapter 10: The Bulgarian Incident

As I mentioned before, I was always "that guy" who was involved in whatever was happening. On this particular night a lot happened, and I found myself smack in the middle of an international incident that could have fundamentally changed the course of the war in Kosovo.

On that particular night I was leading a two-ship, callsign Sword 11, conducting a presence mission over Kosovo. The weather was scheduled to be really shitty, so the NATO CAOC had cancelled all of the missions for the night across the whole of Yugoslavia except for ours. Our job was to roam around providing the Yugoslavians another reminder that we owned the airspace over their country, even in bad weather. Providing presence in bad weather was not as easy as it sounds. My wingman, Sword 12, and I would have to be very careful and stay out of the clouds so that we could see if someone was shooting at us. If needed, we could fly really high to stay above them, but if we couldn't keep ourselves in the clear we would abort the mission and come home.

It was raining pretty hard when we went out to the jets. I always felt really bad for my crew chiefs on nights like this. Yes, I got wet, but things weren't too bad once I climbed in the cockpit and shut the canopy. The F-16 has a great environmental control system, and I felt guilty sitting there all toasty and drying off, while they were moving around my jet in the chilly, dreary European night and getting soaked to the bone. It might have been worth it if we had a real mission of any value that night, but we didn't. The crew chiefs are really an amazing bunch of young men and women to work in the heat, and the cold, and the rain, to give us safe jets to fly.

The weather was really nasty too. Sword 12 stayed in trail about a mile behind me all the way to the tanker. Rejoining on the tanker was pretty tough, with the in-flight visibility so bad that I couldn't actually see the tanker until I was well inside of 100 feet from him. You might not believe me but flying up on another airplane that close without being able to see him always made me sweat more than getting shot at. Between the pitch-black night, the rain, the thick clouds, and the light to moderate turbulence, getting gas was quite a chore. When we were gassed up and clear of the tanker, I breathed a sigh of relief. I told Sword 12 to stay 3-5 miles behind me, and that we would stay that way while we explored the weather over Kosovo.

Much to my surprise and delight, as we crossed the mountains from Albania and into Kosovo, the weather cleared up. And I'm not talking just a little…there wasn't a cloud in the sky, and the air was so clear that the visibility was phenomenal. So much for the weather forecast! I decided that we would still stay in our trail formation, and I climbed us up to a little over 30,000 feet. It was a really beautiful night. We could see the wall of clouds away to the west and south over the Adriatic Sea and Albania, occasionally glowing internally with a flash of lightning. To the north you could actually see the glow of Belgrade's city lights cresting the horizon. East, across Kosovo and into Bulgaria, there was a soft, thin layer of fog blanketing the low places in the ground. It was absolutely amazing to behold.

Leveling off at 30,000 feet or so, I called the AWACS and passed along an update of the weather. I told them that we would fly north along the western edge of Yugoslavia for about half the length of the country, then cut across to the eastern edge and fly back south along the border with

217

Bulgaria. A little math told me that this route would take almost the whole 45 minutes we were scheduled to be on station. After flying that big loop, we would go back to the tanker, get gas, and fly back to Aviano. It looked like it was going to be an easy night.

As we cruised along the radios were silent, except for the occasional "picture clear" call from the AWACS. There was nothing showing on my radar scope either, and my different sensors were not detecting any radar signals. It really was turning out to be a quiet night, and as we reached the eastern side of Yugoslavia, I turned us south along the Bulgarian border. Bulgaria was a formal member of the Warsaw Pact, and they were staying neutral in this fight. There was a rumor in the press that Bulgaria was even trying to get admitted to NATO. It occurred to me that I really didn't know much about Bulgaria.

I was really enjoying the flight. The border between Yugoslavia and Bulgaria was pretty unpopulated. The countryside was mostly dark, but there was one well-lit and sizable Yugoslavian city near the border: Nis. The city of Nis had not played much of a role in the air campaign to date. It was located on the far side the country from Kosovo, and well away from Pristina and the areas where the ethnic cleansing was most prominent. It was also well away from Belgrade and any of the other major target areas we serviced. So, as a result, to my knowledge, we hadn't really flown many missions near it. Looking east from the city, you could see the line of lights where the border crossing into Bulgaria was located. The whole city was covered in the thin fog, which made everything seem to glow. It reminded me of one of those snow-covered city models you see under Christmas trees sometimes, with the buildings nestled in cotton to simulate snow.

I really didn't pay it all too much attention. Everything was so quiet, and at 30,000 feet the stars were very bright, so I decided to watch them instead. I had the autopilot set, my arms resting up on the edges of the canopy, and I tilted my head back to look for shooting stars. I realized that I would have to be careful and not fall asleep stargazing in my warm and cozy cockpit. Relaxing there, staring into the night sky, I physically jumped against my straps as a missile flew across the top of my jet from behind and detonated right above me!

A bucket-full of gourmet adrenaline surged through my body as I grabbed the flight controls. I didn't have time to disengage the autopilot, and that set off warning lights and sounds as the flight control system protested my actions. I rolled my F-16 fast to the left and broke into a high-G turn, slamming the throttle into afterburner at the same time. I twisted around to check my six and see who had shot at me, just in time to see another missile fly by, also detonating as it passed my jet. "Sword 11 flight, break left!" I shouted over the radio. I couldn't see my wingman in the dark with his lights out, and it occurred to me that we would have to depend on the "big sky" theory to avoid hitting him. The "big sky" theory says that the sky is a big place, so the odds are small of hitting another airplane. While not the best plan, it would have to do for now.

Rolling out of my break turn, I dove for a few thousand feet to get my airspeed back up, then cancelled the afterburner so that I would disappear again into the darkness. Searching the sky and ground I couldn't see any more missiles. My mind was racing as I weaved my jet around looking for anyone else shooting at me. I had no indications on my RWR that a radar was locked onto me. I was too high for any heat-seeking MANPADS to have taken those shots. And there was no AAA anywhere. But those missiles came close to hitting me…so close that I could vividly remember seeing small sparks trailing out the back of both missile's rocket motors before they were detonated by their proximity fuses. They had only missed me by a few yards! My mind jumped to a MiG…had a

MiG snuck up on us with his radar off and taken a shot with an air-to-air heat seeking missile? Holy shit!

"Sword 11 is naked" I called over the radio, indicating that nobody had an air-to-air radar locked onto me. Sword 12 replied that he was naked as well. To make sure he understood what I was thinking, I also threw out a "Watch for MiGs" call. Next, I called the AWACS, "Magic, Sword 11, picture at our location!" Almost like he was waking up from a nap, the AWACS controller replied in a non-specific European accent, "Sword 11, picture clear." That meant he didn't see anybody. While certainly helpful, I honestly didn't trust the sleepy NATO AWACS guy with my life right at that moment. I needed to get Sword 12 back into some sort of tactical formation so that we could better protect one other, and to also avoid accidentally shooting each other if there was a MiG mixed in with us. "Sword 11 flight, flow north and take it down (to a lower altitude). Sword 12 say your position." Sword 12 acknowledged, and the position he gave me was only a couple of miles from my own. With a little back and forth over the radio we got organized, and he quickly called "saddled," meaning that he was back in formation.

After confirming again that we were both naked, I turned my attention to looking for MiGs and SAMs. I set us up into an orbit and went around the racetrack a couple of times until I was sure there were no MiGs near us. Confident that we were alone in the sky, I focused on figuring out where the missiles could have come from. The missiles had come from roughly just outside the western edge of Nis. Focusing my sensors on that area I didn't find any SAM radars, but I did find a Soviet-made early warning radar. Because of their long range and how the enemy used them in their integrated air defense system, we rarely found one radiating so close to us. The ROE allowed me to engage it if I wanted to, and after getting shot at I was certainly in the mood for some payback.

The tactical problem I faced was the type of radar I was looking at, as well as its location. Yugoslavia and Bulgaria both were equipped with the same radars made in the former Soviet Union. The location of the radar I had found was awfully close to the border on the Yugoslavian side of the fence, close enough that from the angle I was looking at it, I couldn't be 100% sure which country it was in. To avoid shooting a Bulgarian radar I needed a way to make sure. After thinking about it for a few seconds, I decided that since I could clearly see where the border was by the lights on the ground, I would simply swing into Bulgarian airspace and fly back toward the radar. When I flew across the clearly visible border between the two countries, if the radar was still in front of me that would mean it was in Yugoslavia for sure, and I would shoot it. As I swung our flight eastward, I was pretty sure that at 0100 in the morning nobody in Bulgaria would notice, nor care if they did. And if they did care, they couldn't do anything to stop me…super-power coming through!

Once I had the border comfortably behind me, I turned us back toward Yugoslavia and concentrated on the radar I wanted to destroy. In short order the line of lights on the ground marking the border slid under my jet. The early warning radar was indeed in front of me! I glanced over to make sure that my weapons were armed and that I had a HARM selected. The radar was not very far in front of me, so I pushed the stick forward, the negative Gs floating me up against my straps, and causing all kinds of dirt to fly up from the cockpit and against the canopy. The targeting box around the radar settled into my HUD and I relaxed the stick back to positive Gs…I was now in a steep dive toward the radar and the city that surrounded it. My thumb stroked the pickle button on the stick, and the HARM leapt off my jet like a hound after a fox! I watched as the missile flew straight to its target, looking like a meteorite as it flew, and its rocket engine was still burning when it hit. The warhead on a HARM isn't huge by any means, but it is pretty powerful none the less. The flash of light as it detonated seemed amplified by the thin, soft layer of fog around the target. It was very satisfying to see! After recovering from my dive, I set us up in an orbit nearby to see what the Yugoslavian's response would be, but there wasn't one. All remained quiet. They didn't even shoot a single round of AAA into the sky.

Our gas was soon low, so we went to the tanker. The weather was still shitty while we were getting gas, and we were in the soup the whole way home too. It was a stark change from the clear skies over Yugoslavia. We split up close to the base and flew instrument approaches to landing. If anything, it was raining harder now than when we took off a few hours ago. After we landed and taxied clear of the runway, the first thing we did was stop and some crew chiefs came out to de-arm the airplane. This involved them putting safety pins into all of our weapons so that we didn't accidentally drop them or shoot them off on the ground. De-arming is almost always done by hand-signals alone, especially on rainy nights when the ground crews were in a hurry to get back inside their shelter. Since it was rare to have them take the time to hook up a headset to the jet's intercom system, you can imagine my surprise when the lead de-arm guy checked in with me! "Hey Sir!" he yelled over the engine noise. "It looks like you guys both saw some action tonight!" My thoughts immediately jumped to the near-misses I had with those missiles, and it suddenly occurred to me that maybe I had some damage to my jet that I didn't know about! Trying to play it cool (fighter pilots have a reputation to maintain, after all), I asked him what made him think that? He squinted up at me through the driving rain and said, "Because you both shot a HARM." I thought about that for a second. I knew I had shot a HARM, but when did my wingman shoot one? I turned around and looked at his jet. I could barely make it out through the glare of the stadium lights reflecting in the heavy rain, but sure enough, he was missing one of his two HARMS! I chatted with the crew chief while they finished de-arming me, and then wished him a good night. Taxing back to parking I was kind of pissed at my wingman, because he should have let me know he shot a HARM. Maybe he shot it in the heat of things and forgot to say anything on the radio...kind of like "buck fever" in the moment. Regardless, I would speak to him about it. He needed to do better next time.

Once he climbed into the crew van the first thing he said was, "I need to talk to the Commander." I looked at him.

He was visibly shaken. "What about?" I asked. "I just need to talk to the boss" was his reply. "Okay," I replied, "in the meantime let's talk about the HARM you shot. When did that happen? I didn't hear you call it on the radio." He didn't look at me. Instead, he just repeated again, "I need to talk to the Commander." This was really odd, and we rode back to the squadron in silence. As soon as the van stopped, he was out the door and off like a shot. As I watched him go in search of the boss, I wondered what could have happened that spooked him so badly? And how did shooting a HARM play into it? For a second, I thought maybe he had shot a HARM at me somehow, but that didn't make sense since he only fired one HARM and I had two missiles blow up next to me. I decided not to worry about it for the time being, and I was quite sure that if something bad had happened I would hear about it eventually.

After dropping off my flight gear and debriefing with our Intelligence people, I looked at my watch. It was a little after 4am. I was hungry and the chow hall wasn't going to open until 0500, so I decided to take advantage of the late hour and find a computer to send a few emails home. Normally it was hard to find an unused computer, but at this time of the morning it was no problem. I logged in and checked my work emails, and then spent an hour or so sending personal emails to family and friends. In 1999 having email at a deployed location was a real treat. Back when I deployed to Operations Desert Shield and Desert Storm back in 1990 through 1991, we had to send letters the old-fashioned way. It took about a month for a letter to get back to the States, a reply to be written, and then sent back. Some letters took longer than others too, so it wasn't uncommon to receive letters out of sequence. That was sometimes very confusing! After taking care of emails, I still had a little time to kill, so I decided to check the news online. As the website loaded, my eyes were drawn to an article titled, "Errant NATO Missile Strikes Apartment Building in Bulgarian Capital." I leaned forward and made sure I read that right. As I waited for the story to open, I wondered if this was connected to Sword 12's wanting to find the Commander so badly? I had a gut feeling it was.

As it turned out, the errant NATO missile was indeed the HARM my wingman had shot without telling me. It was all over the news the next day. I believe Stars and Stripes even carried a picture on the front page of their daily edition showing the President of Bulgaria standing in the rubble of an apartment with the wall and roof missing, looking at the shredded furniture with his hands on his hips.

Fortunately, nobody was hurt, but the apartment building was in bad shape. And, of course, it was a huge international incident. Fortunately, the Bulgarians didn't go too crazy over it. A few years later, I found out why.

In the meantime, there was an investigation. And I have to commend the Air Force for how well they conducted it. A full Colonel from the Pentagon was sent to look into it. The guy was a former Electronic Warfare Officer who had spent most of his career flying F-4G Wild Weasels and had flown combat missions in Desert Storm. That was perfect, because his technical background meant that he genuinely understood everything he was looking at, and his combat experiences allowed him to put it in context. Here is what he figured out. When the two missiles flew by my jet, Sword 12 initiated his own break-turn before I even called for it on the radio. In the middle of that high-G break turn he looked down at his sensors and saw a radar, so he quickly shot a HARM at it. The problem was it was a radar deep inside Bulgaria. The missile tried its best to get there, but eventually ran out of speed and fell to the ground, right into an apartment building in downtown Sophia. The missile's warhead did not detonate as it would have done on a target. Instead, the impact of hitting the building set off a low-order detonation. That was a stroke of luck and probably was the reason nobody got hurt. Sword 12 was under attack at the time he took the shot, and while he should have determined where the radar was before shooting, it was reasonable that he returned fire in that moment. NATO followed the Colonel's assessment that this was an accident caused by the fog of war, and not because of negligence or ill intent. We were all pleased when the word came down that Sword 12 wasn't in any trouble. We were even more pleased that the U.S. or NATO didn't put some stupid restriction into effect on how we could operate, in a misguided effort to prevent it from happening again.

I mentioned earlier that a few years down the road I found out why the Bulgarians didn't get more upset. Fast forward to 2002, and I was reporting to Air Command and Staff College (ACSC) at Maxwell AFB. ACSC is a 10-month long Air Force school for Majors to learn about how to plan air wars, and how to be effective squadron commanders if they ever got the chance. The class size each year is pretty big, with about 800 students attending. The class was made up of not only Air Force Majors, but officers and civilians of equivalent rank from the other branches of the U.S. armed forces, and something like 30 or 40 foreign countries. The student body was divided into clusters of 12 or so students, called Seminars. Each Seminar contained a cross section of folks from the USAF, the Army and Navy, and a couple of foreign officers. As I sat down with my Seminar on the first morning, I discovered that the foreign exchange officer sitting next to me was from Bulgaria! It never occurred to me that there might be a connection between that HARM hitting the apartment building and the guy sitting next to me. He was a pretty interesting guy, and he contributed a lot to our different classes. But one day, our homework included readings on the NATO air war over Kosovo. The instructor led our Seminar though a lesson on the challenges and complexities of operating in a multi-national coalition like NATO. Many of the folks, myself included, had been there and a lively discussion ensued. At one point, the Bulgarian guy spoke up. And here is the story he told.

Late one night he was sitting on duty in the Bulgarian Air Defense Headquarters operations center. It was a quiet night, and they were watching two American F-16s flying around near the city of Nis on their border. All of a sudden, one of the F-16s shot a missile into Bulgaria, and soon thereafter the two F-16s invaded Bulgarian airspace. Their General who was in charge that night was certain that this was the opening move of a NATO attack on his country, and he ordered the Bulgarian air defenses to high alert. While he was getting the President of Bulgaria on the phone, he ordered one of their long-range SAM systems, an SA-10, to prepare to engage us. As he was explaining to the Bulgarian President what was going on, reports began to roll in of the missile hitting an apartment building in the capital city. The General then noticed that the two F-16s were retreating back toward Yugoslavian airspace and requested permission to shoot them down before they escaped. To the General's amazement, the Bulgarian President told him not to shoot. The President then explained that Bulgaria was seeking admission into NATO, and that acting with restraint would give Bulgaria tremendous political leverage during the coming entrance talks. The General hung up the phone, and with obvious disappointment ordered everyone to stand-down.

Needless to say, my jaw dropped. What a coincidence to be sitting next to this guy and hearing another side of that story. It was especially eye-opening to learn that an SA-10 had targeted me and my wingman! At that time the SA-10 was probably the most lethal Russian-made SAM in existence. If they had decided to attack us with it, we would have probably been blasted out of the sky without ever seeing it coming. That story also explained why Bulgaria didn't go off the deep end over the incident. And I guess their President's restraint paid off, because in 2002 Bulgaria officially became a member of NATO.

Chapter 11: Bounty Hunter

Not every mission that was fun and exciting involved me dodging SAMs in the middle of the night. One beautiful day I found myself on the schedule leading a two-ship to the KEZ, callsign Postal 21. We didn't have anybody bombing a target to support; we were just going there to provide presence while the peace talks to end the shooting were wrapping up. Over the last few days, the pace of operations had really slowed down, and most of the sorties were pretty uneventful.

Postal 22 and I met up for an early lunch, and after a quick briefing we stepped out to the jets. The weather was absolutely gorgeous, and as we transited down the Adriatic coast of Italy, I took the time to really enjoy the scenery; I figured it could be one of my last flights in this war. It was also a nice change of pace, since for the last couple of months the flying was intense, and instead of enjoying the view on the way to the tanker I was focused on the details of my mission...or trying to get my INS to work! At the designated point, we turned east and crossed the Adriatic Sea for Albania. The airspace over Albania was where the majority of the air refueling tankers had worked since the war started. At the height of the war the air refueling tracks were stacked with upwards of a dozen KC-135s and KC-10s at the same time. The sky was crowded with fighters coming and going, trying to locate the correct tanker to get their gas. When the weather was bad, especially at night, the air refueling tracks were downright dangerous and you had to be on your game to not run into someone else in all of the chaos. But today there were only a couple of tankers on station, and we were the only fighters.

Everything was very leisurely as we departed the tanker track and headed over toward the KEZ. Along the way I kept staring at the incredibly rugged and beautiful mountains and valleys below us. The more I stared, the more I couldn't stand it. I needed to go fly low-level through them! It was against the rules, but I was willing to take the chance. We had already fenced-in, so our transponders were off which meant that the AWACS wouldn't get an automatic readout of our altitude. Since it was a NATO AWACS, I was quite sure they wouldn't care if they did notice us having some fun. Over our inter-flight radio, I simply clicked the microphone twice to get my wingman's attention. Then I rolled the jet over on its back and dove for the ground with Postal 22 in hot pursuit!

I have always loved flying low-level through the mountains. Deep in a valley, with steep cliffs and brilliantly sun-lit tree-covered slopes close on each side, the sensation of speed is amazing. Streaking along at 400 knots, twisting, and turning to follow the valley floor, I felt like Luke Skywalker flying in that trench to destroy the Death Star! I picked our way through the mountain passes to keep us going toward the KEZ, and in a few minutes we were through and over Yugoslavian territory. We zoomed our jets back skyward, leveling off around 20,000 feet with a series of aileron and barrel rolls to finish it off. Postal 22 was close enough off my right wing that I could see him clearly in his cockpit. He held up his fists and made like he was boxing. I nodded vigorously back at him, and then rapidly pushed my stick forward and backward several times. This made my jet bob up and down like a porpoise, which was the visual signal for him to move out into tactical formation, about 9,000 away and a couple of thousand feet higher. "Damn!" I thought as I watched him bank sharply away, "I can't believe we get paid to do this!"

Now that we were done goofing off, it was time to get down to business. I constantly reminded myself that we couldn't get complacent. Peace might break out at any minute, but until then there were still people here who wanted to kill us. With that in mind I very diligently worked my radar and sensors, and visually searched the sky and ground for threats. So, for the next 45 minutes the plan was to fly back and forth in a north-south orbit about half-way between the tanker track and Pristina, kind of near the Macedonian border. After just a few minutes, I began to notice that there was an early warning radar to the north of us, as well as one to the south. Whenever we turned north, the radar to the north would shut down, and the one behind us to the south would begin radiating. And when we turned south, they would swap, with the southern radar going silent and the one to the north coming back up. This captured my attention, as it was obvious that they thought we couldn't see them behind us. As we flew back and forth in our orbit, I puzzled over their tactic, trying to figure what they were up to. Kosovo was pretty much cleared of radar guided SAMs by this point on the war. The mobile SAMs in other parts of the country had largely gone to ground and were in hiding, since every time they showed themselves, we blew them up. The Yugoslavians still had plenty of

MANPADS, but we were high enough that those weren't a realistic threat. So, when I added all of that up, I ruled out that the early warning radars were trying to catch us in some sort of "SAM-bush." I also considered that maybe they were looking to sneak a MiG up on us, but however unlikely that might be, I made sure to keep a close eye on my air-to-air radar, checked in frequently with the AWACS, and did a good job of visually checking six. This cooperative blinking of the early warning radars kept up for the whole 45 minutes we were on station. Eventually, we were out of gas and needed to go back to the tanker track for more.

While we were going to the tanker, I kept trying to figure out what the Yugoslavians were up to. By this point in the war, they were very familiar with our flight patterns and routines over Kosovo. Perhaps our low-level earlier, which was highly irregular, had piqued their curiosity and they were thinking that we were up to something special?

After getting more gas we headed back to the KEZ. I decided that we would do a repeat of our low-level, and with a double-click of the microphone, Postal 22 followed me down and through the mountains again. Popping back up to 20,000 feet once we entered the KEZ, I smiled in the satisfaction that it was as much fun to do that the second time as the first!

And just like the last time, the two early warning radars started playing their cooperative blinking game, trying to watch us without being noticed. It really began to bug me, trying to figure out what they might be up to. Both radars were in range of my HARMs, the one to the north just barely. The ROE certainly allowed me to shoot them, but I was very skittish about shooting the one to the south. It was right on the border with both Albania and Macedonia, and the drama surrounding my Bulgarian adventure was a cautionary tale fresh on my mind. The last thing I wanted to do was take a chance on sending a HARM sailing off into another friendly country. So, I began to think in earnest about popping the northern one.

The problem with shooting the northern radar was that he kept turning his radar off whenever I pointed his way. Given the long range, to take a better shot I would have to fly closer, but that would definitely tip him off that I was on to him. I also figured that the Yugoslavians were monitoring our radios, so if I tried to explain to Postal 22 what I was up to, that would blow my cover as well. I thought about it for a while, and eventually decided on a way to pull an attack off without signaling what I was up to.

I wanted to keep lulling them into a false sense of confidence that all was well, business as usual, and we didn't see them. So, for our whole 45 minutes on station I kept us in our orbit, turning north and south as regular as clockwork. At the end of our time on station I began turning us to the south from the north end of our orbit and called the AWACS letting them know that we were done and heading home. At the same time, I was double checking that my HARM was set up and targeted for that northern early warning radar. Just as we rolled out southbound, and without any warning to Postal 22, I jammed the throttle into afterburner and wracked my F-16 into a high-G turn back north. Pausing for only a second to make sure I was pointed at the northern radar who was now radiating, I launched a HARM. As soon as the weapon rocketed away from my jet, I ripped into another high-G turn back to a southern heading. I am sure my abrupt maneuver surprised Postal 22, but I was hoping that it all happened so quick the radar operators wouldn't notice the momentary deviation from a predictable turn to the south. Looking over my shoulder, I could see the HARM rocketing away and climbing, leaving a contrail as it arced through the stratosphere.

I never said a word, and neither did Postal 22. After getting gas from a KC-135, we headed for Aviano. The trip back to base would take us up the Adriatic coast of Albania, Montenegro, a tiny bit of Bosnia-Herzegovina, and then along the coastal islands of Croatia. It was another beautiful flight, with the sun reflecting off the blue water and the waves breaking on the beaches. The Croatian coast and islands were especially gorgeous; I had heard the resorts there were really nice too. I also remembered watching a TV show where they talked about the waters off Croatia being a breeding ground for Great White sharks. If I ever visited one of those islands, I would have to remember to limit my water sports to the pool!

Back on the ground Postal 22 and I debriefed Intelligence on the radars and my HARM shot. We went over to the Officer's Club for steak dinners and chatted about the mission. He too had been watching the early warning radars, and when he saw me spin around as we were leaving, he knew what I was up to. A few days later when I was in the squadron, one of the intelligence officers pulled me aside and shook my hand. "Congrats, man, we received word this morning that your HARM hit that early warning radar! That was a one-in-a-million shot!" I thanked him for the compliment! A few days after that I was getting ready to step out of the squadron to go fly, when a pilot from the F-117 squadron walked in with a brown paper bag in his hand. He asked us if Captain Littleton was around and grinned from ear to ear when I raised my hand and said that was me. "The boys over at the squadron wanted to show our appreciation for bagging that early warning radar. We hate those things, but nobody ever shoots them. Here, this is for you," he continued, as he pulled a fifth of Jack Daniels out of the paper bag and handed it to me. "We had joked that maybe more Wild Weasels would shoot them if we put a price on those early warning radar guys' heads. Consider this a bounty payment from the Nighthawks." I shook his hand and thanked him back, adding, "I guess that means I can add 'Bounty Hunter' to my resume now!"

That was a really cool gesture from the F-117 guys. I never opened that bottle, and to this day it sits proudly and prominently on a shelf in my living room with all of my other "I love me" stuff. When people visit and look at the pictures on the wall, and plaques and stuff, I almost always get asked, "What's up with the bottle of Jack?" And I reply, "Well, there's a pretty good story about that!"

Chapter 12: Tanker Track Excitement

One evening I found myself leading a two-ship over the KEZ, callsign Spear 21. The mission had actually started during the day, with Spear 22 and I taking off in the middle of the afternoon. The weather was pretty good, with scattered clouds across Italy, the Adriatic Sea, and Kosovo. As usual, we refueled in the tanker tracks over Albania, and then proceeded into the KEZ where we provided support to a flight of A-10s who were looking for targets worth striking. Searching for small, hidden, or mobile targets was a primary mission in Kosovo, and I was hopeful that the A-10s would find something interesting and worthwhile.

Finding previously unknown targets, identifying them, and killing them is a lot harder than you might think. The various wars on terrorism since September 11th, 2001, have largely involved this kind of work, finding, identifying, tracking, and striking mobile or hidden small targets. In the last couple of decades, the U.S. military has developed marvelous technologies and sensors to refine that process. We make it look deceptively easy now…kind of like air-to-air. But back in 1999, during the Kosovo air war, we weren't nearly as advanced. We were still pretty good at it, but it was a lot of work, and the fact that the ROE didn't allow anyone to fly below 15,000 except in an emergency or to actually deliver a weapon, made it a lot harder. Believe it or not, the A-10s didn't have any kind of electro-optical sensors to search with; they were using binoculars and their own eyeballs. The F-16 FAC-As didn't have it much better; in addition to binoculars, they had LANTIRN targeting pods, but those early generation pods didn't have the resolution or magnification for effectively identifying what an object on the ground was except under the most perfect of circumstances. After the Kosovo war was over, the Yugoslavians openly bragged in the media that NATO had been tricked into wasting a lot of bombs on dummy tanks and other equipment that they made out of wood and plastic sheeting. You can blame a lot of that on the restrictive ROE our hunters and killers had to deal with.

Just a few days before my mission, one of the guys from another F-16 squadron found himself over the KEZ searching for targets of opportunity. It was night, so his binoculars were of no use, and he was stuck searching up and down roads and in fields with his LANTIRN pod. Searching large areas with a targeting pod is hard to do; the field of view for a targeting pod is only a couple of degrees wide. It's like looking through a soda-straw. It's one thing to know roughly where a target is and find it with a targeting pod, but it's almost impossible to find an unknown target just by sweeping around with a soda-straw field view of the world below. But that was what the mission was, and this F-16 pilot was doing his best, when he stumbled on something significant.

While searching up and down a remote country road that wound through some farm fields and woods, he spotted a dozen "hotspots" in a line. LANTIRN pods see the world in terms of thermal contrast and turns it into a picture based on only the contrast it can differentiate between hot and cold. So, the "hotspots" he could see indicated something much warmer than the area around it. The resolution on his LANTIRN pod wasn't very good, not only due to the inherent limitations of the pod itself, but due to the high humidity which made the picture fuzzy and less distinct. But from experience, he knew that hotspots like this were exactly what the engine compartments on tanks and other armored vehicles looked like. As he watched the hotspots proceed down this country road, he felt like he might have hit the jackpot! But in accordance with the ROE, he needed to complete a fairly complex matrix of identification procedures to rule out that it was a convoy of civilians or other off-limits targets. This process took a while and required extensive communications and coordination with the NATO command and control structure. Finally, he received the word; there were no friendly fire concerns or collateral damage issues…he was cleared to kill those tanks!

He quickly called the AWACS and asked for them to send some of the fighters that were nearby, a flight of F-15E Strike Eagles. The F-15Es carried a lot of laser-guided bombs (LGBs), and he could direct them so as to strike each and every tank. While they were on the way, our intrepid F-16 pilot decided to use the two LGBs that he was carrying. Thinking through how to best attack the tanks, he came up with a plan. He would bomb the tank at the front of the column first to block the road, and then he would bomb the tank at the back of the convoy so that the rest of the tanks couldn't turn around and escape that way. Locking his targeting pod onto the lead tank's hotspot, he released a single 500-pound LGB, called a GBU-12, and moments later scored a direct hit. As he watched through his LANTIRN pod, he could see the hotspots of the remaining tank engines stop moving, and then reverse direction back down the road, just as he predicted. Having correctly anticipated that this would happen, he was ready and soon dropped his second GBU-12. Arcing through the night sky, he guided it right onto what was now the new lead tank driving in the opposite direction. The effect was spectacular, and in a panic the remaining tanks began to scatter in all directions into the fields and trees along the road. The F-15Es arrived about then, and for the next half hour the F-16 FAC-A coordinated the bombing of each and every hotspot as is tried to run away and escape destruction. After all of the tanks had been destroyed, the F-16 pilot returned triumphantly to Aviano, and basked in the glory of his work that night.

But that wasn't the end of the story. A few days later a U.S. Special Forces team was in the area, and they were tasked to check out the bombed tanks and gather any intelligence they could find. It turned out that the dozen hotspots that were bombed were not the heated engine compartments of tanks, but rather the warm bodies of a herd of cows! Picture that in your mind; one moment you and your fellow cows are walking down a dirt road on a nice cool night, looking for a good place to eat some grass, when suddenly the cow leading the herd is blown into bits by a 500lbs bomb. Shocked and frightened, you and your herd-mates turn around and run the other way, only to see the new lead-cow vaporize in a blinding flash of fire and thunder. Everyone now panics, and the integrity of the herd is lost...it's every cow for him (or her) self! Chaos reigns, and cows are running in every direction, and the panic grows as one by one your herd-mates are obliterated. But no matter how far you run, you begin to realize you will only die tired, so you decide, "What the hell, I will just stop and eat some grass and think happy cow-thoughts until my time comes."

Word of what really got bombed spread like wildfire! It was one of the funniest things ever to happen while flying fast-jets, and to this day I still chuckle when I think of it! I can only imagine the shit that F-16 pilot took from his fellow pilots! In the aftermath, the jokes and comments were endless. One of the best was pulled-off by his squadron's intelligence shop. The intelligence officers in a squadron work hard to help educate pilots on the different threats they may face. This mostly occurs via formal briefings in different venues. But one creative way they often use, is to hang information over the urinals in the men's room, so that while guys are taking a piss, they can learn something of value instead of just admiring themselves. Obviously, you don't stand at the pisser for too long, so the information has to be brief and to the point. Intelligence officers also know that fighter pilots like pictures, so there is usually a picture of the threat as well. A good example might be a picture of a MiG-29, with a description of what weapons it carries and how fast it can fly. So, to make fun of the cow-killer, his intelligence officers hung a new "threat" page over the urinals. It had a picture of a cow with the title, "Species: Bos Taurus Linnaeus, more commonly known as the Domestic Cow." Under the picture were listed some facts every pilot should know:

"Description: four legs, black and white camouflage pattern, all-terrain capability, some models equipped with udders.

Armament: Four hooves. Some models equipped with horns.

Top Speed: 3mph, with short bursts of speed up to 10mph when herd-mates explode."

I never found out exactly who that unfortunate F-16 pilot was, but if I ever meet him, I will buy him a steak dinner and a glass of milk to wash it down!

On this sortie, however, nothing so funny or interesting was happening. In fact, the A-10s found nothing at all of interest, and Spear 12 and I enjoyed another afternoon of making left-hand turns over the Kosovo countryside. We stayed on station for 45 minutes, air refueled, and did it again for another 45 minutes; rinse, lather, repeat. As our last time on station expired, and it was time to head back to Aviano. Spear 12 and I went back to the tanker track over Albania, got just enough gas to return to Aviano, and were soon on our way back up the Adriatic. About five minutes after we left the refueling tracks, I was watching the sun's last rays nip below the horizon, when the AWACS called us. "Spear 11, Magic. We need you to turn around...the tankers are under attack!"

That jolted me from my relaxed sightseeing in a fast second. "Spear 11 copies" I replied, "Spear 11 hook right, Fence-In." As Spear 12 and I made a hard 180-degree right turn, my hands flew about the cockpit as I armed my weapons and tried to think of what could be happening. The air refueling tracks where comfortably outside of hostile airspace and over the mountains and coast of Albania, so I didn't think a SAM could be shooting at them. It must be MiGs! It made sense; KC-135s and KC-10s are pretty defenseless but highly important. Sneaking one or more MiGs underneath them wouldn't be hard to do by having them fly low through the mountains, where they could pop up and attack without any warning. It was time to talk to the AWACS to get a description of where the MiGs were. But when I queried, Magic had a totally different story to tell. There were no MiGs. The tankers were getting shot at by SAMS, but no one had been hit, and the tankers were all running away to the east, recognizing that the airspace over the Adriatic Sea would offer safety.

Spear 12 and I began looking on our sensors, but there were no radars anywhere to be seen. "It must have been a MANPADS launch," I thought to myself. That made some sense, given that if someone carried a MANPADS up on top of one of the mountains under the tanker tracks, they could easily shoot high enough to hit a tanker. I kind of wiggled in my ejection seat at the possibility of this, and if true how it would drive a major change in the conduct of our operations. But there would be plenty of time to worry about that later, right now I needed to figure out where these guys were and kill them.

As we approached the gaggle of tankers over the coast of the Adriatic, I switched Spear 12 and I over to their frequency. Checking in, I asked which of the tankers had been shot at. One tanker spoke up, so I asked him to tell me exactly where he was when it happened, and what he saw in great detail. Apparently, he was in one of the orbits at 20,000 feet, when out of the darkness below a huge rocket came flying up. It passed in front of him, not too close, and kept going higher, and then flew off to the northeast. I asked if there was more than one missile launched, and he said no. As soon as he finished speaking lots of other tanker pilots excitedly chimed in that they had seen it too, each adding their details. From the sound of their voices, you could tell the tanker guys were really rattled, and I couldn't blame them. I was really puzzled by the fact that the missile looked big and kept going, flying out of sight into the distance. That really didn't sound like a MANPADS at all. I chalked it up to the fact that tanker guys don't ever see SAMs, so maybe that influenced their perceptions of what happened. Not everyone makes a living getting shot at by SAMs like Wild Weasels do.

I needed to decide what was the best solution, but to come up with a plan and execute it, Spear 12 and I needed more gas; earlier we had only taken enough to get home comfortably. If we were going to hunt around for SAMs late into the night, we would need to top our tanks if possible. Tankers usually don't have any extra gas to give, and actually get kind of rude sometimes in saying no, so I wasn't expecting any help. But it never hurts to ask, and I figured that worst case I would stay with them as long as I could and then divert with Spear 12 into one of the Italian airbases just across the Adriatic from Albania. We could get gas there and then fly back to Aviano while the NATO commanders freaked out over what had happened. I called out over the tanker frequency, "Spear 11 and 12 need some more gas, does anyone have some extra they can offload?" The response was overwhelming, and the radios were jammed as several tankers all offered up as much gas as we wanted at the same time. I shook my head and laughed, I called Spear 12 over our private inter-flight radio, "They never have extra when *we* need it, but when *they* need us to have it there's plenty to go around!" Spear 12 clicked his microphone twice in humorous acknowledgement.

I quickly came up with a plan and briefed Spear 12, the AWACS, and the tanker, using our encrypted radio capability just in case the Yugoslavians were listening to the tanker radio frequency. I asked the same tanker who was shot at by the SAM to give us the gas we needed, and while refueling us to begin flying back to the tanker track where they were shot at. The AWACS would make unencrypted radio calls like he was clearing the tanker back to work and sending my two-ship back to Aviano. I had the tanker and Spear 12 turn off all of their exterior lights, and I left my lights on so that any observers on the ground would see a single airplane up above. After giving us a full offload of gas, the tanker silently turned around and flew back to the safety of the Adriatic Sea. Spear 12 and I continued on, hoping that whoever had shot the SAM would decide to shoot at us, thinking we were a tanker. Once they shot at us, I would be able to figure out how to best kill him once I knew what kind of SAM it was and where they were shooting from.

I did my best KC-135 imitation once we were at the location of the previous launch. I flew around at 300 knots, following the same racetrack orbit the tankers flew, and making tanker-type radio calls while using the callsign of the tanker who had experienced the SAM launch. Looking down into the darkness, I could just make out the shapes of the mountains below. Spear 12 maintained radio silence as we orbited, both of us intently scanning back and forth between our onboard sensors and the ground below. After a half-hour or so, I decided that nobody was going to take the bait. Changing tactics seemed like a good idea, so with the time and gas we had remaining, Spear 12 turned on his lights, and we descended to a much lower altitude, trolling around all over the area hoping to draw fire. But nothing happened, and with our gas running low, we chatted with the AWACS to let him know we couldn't find anything. The tanker, who was still hanging out nearby, offered us more gas. We really didn't need it, but I decided that having full tanks for the flight home would be a good idea; it had been a weird night and could get weird again.

The flight back to Aviano was pretty unique. Shortly after we left the tanker track for Aviano, I started thinking about how much gas we would still have when we landed. One of the many nice things about flying the Viper is that is can actually land at its maximum gross weight. Many other airplanes can't do that and need to either burn down some fuel or dump it to get to a safe landing weight. But just because you can do something doesn't mean you should do something. Both me and my wingman were very experienced Viper pilots, but why tempt fate with a heavy weight landing when we could burn the fuel and have some fun at the same time! We had been cruising northward up the east side of the Adriatic Sea, and I could see some huge thunderstorms building just inland from the coastline. The cumulonimbus clouds were flickering and glowing with the reflections of lightening popping out of their bottoms. It dawned on me that the combination view of those storms below and the stars above would be spectacular at 50,000 feet. So, with a grin, I cleared my wingman back into two-mile trail, and lit the afterburner! The jet quickly accelerated, and at Mach 1.2, I slowly raised the nose to hold the climb at that speed. The altimeter began to spin fast as we rocketed into the night sky. I twisted around in my seat and marveled at the glow of my afterburner plume, and I could also see Spear 12's burner plume behind and slightly below. As we passed through 45,000 feet, I began easing the nose down, and backing off the throttle a bit. I leveled us off at 50,000 and set the throttle to stay at Mach 1.2.

Looking around from that altitude is an amazing experience. In daytime you can see the curve of the earth, and the sky overhead is actually more black than blue. At night it was a whole different experience. To the west the rim of the earth was a fiery line, still ever so slightly aglow with the setting sun just below the horizon. Looking up and to the east, the sky was filled with stars, and the Milky Way was like a blanket of light. And looking down to the east at the top of the thunderstorms was like nothing I had ever seen before…the clouds looked like popcorn popping with light as the lightening they generated crackled inside. And then I saw something that I had never seen before and have never seen since. A spiderweb of lightening shot back and forth between two adjacent storm clouds, and a second later an enormous bolt of lightning shot out of the top of one of the clouds and up toward space! If I hadn't seen it with my own two eyes, I would never have imagined such a thing could happen…it was so beautiful and fearsome at the same time. The sheer size of the lightning bolt is still burned into my minds-eye today, and after seeing it I was always a little more respectful of the power of mother nature.

Back at Aviano there was a lot of buzz about that SAM launch in the tanker tracks. Spear 12 and I were celebrities and were debriefed several times by intelligence, as well as a couple of the Colonels. The NATO CAOC called us too. The reason this was such a big deal, was that if there were SAMs under the Albanian air refueling tracks, the tankers would have to move to a safer location. That would mean a total restructuring of how the air war over Kosovo was conducted: not an easy task. Disappointed that we didn't find anything, Spear 12 and I answered everyone's questions before leaving to go get some chow and call it a night.

By the next day, when I came back to the squadron, the whole mystery had been solved. The tanker guys had indeed seen a huge missile fly by them, but it wasn't a

SAM. It was an MGM-140 Army Tactical Missile System (ATacMS) fired by the U.S Army into Kosovo at some target. The Army guys didn't think to coordinate the launch from their base in Albania with the NATO CAOC. When asked why, they apparently replied, "Well, an ATacMS missile only goes up to about 20 or 30 thousand feet, and the Air Force guys are way higher than that." I am sure the Army guys were horrified to find out they shot that thing right through all of the KC-135s and KC-10s orbits. Luckily no one was hurt, and in the end, it was another example of the Big Sky Theory at work!

Chapter 13: Work Hard, Play Hard

The Kosovo war wasn't all work and no play. Indeed, I would have to say that of all of my wartime deployments it was the one with the most fun by a long shot. And that's saying something because the Air Force is known for taking care of its pilots! There is an old joke that when the Air Force gets money to build a base somewhere, the first thing they do is build a world-class golf course, a spacious Officer's Club, a large well-equipped gym, beautiful dormitories, and a well-stocked Commissary and Base Exchange. And then the Air Force goes back to Congress and asks for some more money so they can add a runway too! That's just about true. The other branches of the armed service like to make fun of the Air Force and say that we have a soft life. One Army guy I served with used to call us the "Chair Force." My response was, "Hey, there's no draft. Nobody made you join the

Army/Navy/Marines. If you didn't want to sleep in fox holes or spend six months at a time on a boat with 4,000 other men, then you should have made a better choice!" At the beginning of Desert Shield, we stayed with the Army in tents and shitty dorms at a base north of Riyadh, Saudi Arabia. Right before Desert Storm kicked-off, however, we were moved to Jeddah to keep us out of the range of the SCUD missiles Saddam was teeing-up. Amazingly, we were moved into a compound originally build for U.S. contractors working in Saudi Arabia, and the accommodations were first-class! We stayed in three-bedroom villas in a compound by the Red Sea. Every day after flying our combat missions over Iraq, we would don our swim trunks, grab our snorkeling gear, and walk over to the Hilton Red Sea Resort. The Hilton Red Sea Resort had a fantastic beach, alcohol, and great food, so it was a favorite hangout for the non-military expats who lived and worked in Jeddah, Saudi Arabia. In a great display of American patriotism, the Hilton allowed us battle-weary zipper-suited sun gods free access to their beach and facilities. We would swim around and snorkel on the spectacular coral reefs, lay in the sun, and drink beer with the large number of British expat nurses who frequented the Hilton in their off time. War is hell, boys…that's a fact!

Operation Allied Force, the Kosovo Air War, was even better. I was one of two pilots on the team that flew over in advance of the whole squadron arriving, and my job was to arrange billeting and transportation for everybody. When I went over to the base billeting office to get it all arranged, I was informed that all of the hotels nearby were already filled with deployed military personnel from across Europe. The nice lady helping me decided that they would have to send us to a location over an hour's drive away. I told her that was unacceptable, because it would eat too much time out of the pilot's rest schedule. Additionally, I also added that if we needed to bring in extra pilots in a hurry there was no way an hour or more response time would work. The civilian employee went and got the Major who was running the billeting office. We talked about it, and he was sympathetic; my arguments won him over. He said he had an idea, but he would need to get permission, and to hang on for a few minutes. When he returned from talking to his boss, he smiled and said his idea had been approved. Right next to Aviano are the Dolomite Mountains. There was a town about 20 minutes' drive from the base, at a high elevation up in the mountains, called Piancavallo. It was a ski resort town, but because of the steep, narrow, curvy, snow-covered roads winding up to Piancavallo this time of year, they had been prohibited from billeting anyone up there. But since there was no other option to keep us close to the base, the billeting Major's boss gave him special dispensation to put us in one of Piancavallo's hotels. The Major had his Italian civilian staff make some calls, and in about a half hour I was informed that rooms for all 40 of our pilots had been secured at one of the ski resorts. And the rooms came with a free breakfast every morning too. I couldn't believe my luck. This was too good, even for an Air Force deployment! With directions to the ski resort in hand, my next stop was to secure a small fleet of cars for the pilots to drive back and forth between the ski resort and the base.

At the base Transportation Office my luck continued to hold. Every rental car agency for 100 kilometers was out of rental cars. I asked the transportation officer in charge if he had any ideas, since we had to have enough vehicles to get us back and forth from our ski resort to the base. We would need a lot of cars since we would be flying around the clock, and the pilots would need to travel in small groups as a result. He asked me how many cars we needed, and I told him 12; 10 for the pilots, one for the squadron commander, and one for the Operations officer. The transportation officer starred at his desk for a moment, tapping his pencil on the edge, obviously deep in thought. Soon, he picked up the phone and started making calls to the different rental car agencies in the area. He started off asking for 6-9 passenger vans but had no luck. After exhausting his rental car rolodex, he sat back in his chair and stared at me in defeat. With a deep breath, he mumbled something like, "There's no other option" and picked up the phone again. It was all I could do to keep a straight face as I listened to him make arrangements with a local rental car agency for a small fleet of Alpha-Romeo T-Spark sports cars to be delivered to the base ASAP. After getting off the phone, he explained that these Alpha-Romeos had a small backseat, but if we didn't mind being cramped, we could stuff four people in each one. If I couldn't believe my good luck with getting us rooms in a ski resort, you can imagine how incredulous I was with this bit of news! I nodded solemnly, and leaded forward, breaking into a huge grin. "We'll manage! How soon can I pick up them up?" I asked. Two hours later he walked me out to a parking lot and handed me a big manilla envelope with the keys to the 12 sports cars parked there. He left, and as I stood there in front of all these Alpha-Romeos, I looked up at the snow-covered Dolomite Mountains where our ski resort would be. I laughed out loud. "When the guys get here, I am going to be a hero!"

Next, I picked out one of the cars and hopped in to drive up to Piancavallo to get the keys to our rooms. As I pulled out of the front gate at Aviano, I looked at my watch so that I could time how long it took to get to the hotel. That would be useful when I began planning our first few days of operations; I would know what time each flight of pilots needed to leave the ski resort to get to the base on time for their mission.

The drive itself was both beautiful and scary. The roads were very narrow two-lane affairs and made their way up the mountain in a series of long straightaways with blind, hairpin switchbacks. There were no guardrails or shoulders at all, just those short concrete "dragon's teeth" imbedded in the ground along a couple of stretches. In many locations running off the road would plunge you over a several hundred feet high cliff. The roads were snow-packed and icy in many places too. I could see why the billeting folks were hesitant to let anyone stay up here! As I pulled into Piancavallo, I noted the time: 21 minutes. I was sure that the trip back down would be a little quicker, but I would factor in 30 minutes to be safe.

The ski resort was amazing. It was the very stereotype of an alpine resort from a James Bond movie! It literally sat at the foot of the ski slopes, which were still open and thick with skiers. Inside the lodge itself was a good size lobby with a big fireplace, and a front desk with a small bar at the far end. The logs in the fireplace were blazing away and surrounded by people warming themselves after a day on the slopes. Walking in wearing a flight suit I drew a lot of attention, and especially enjoyed the smiles from all of the good-looking European snow-bunnies! I went up to the front desk and was pleasantly surprised that the proprietor was expecting me. This resort was a family owned and operated place, and the staff was very welcoming from the start. They were genuinely nice people, it turned out, although I am sure their good cheer also came from the knowledge that 40 of their rooms would be filled as the ski season ended and spring and summer arrived. I am sure the Air Force was paying top-dollar too.

The owner was an older man we ended up calling "Rocky" because his last name sounded like Graziano. His wife, Anna, was a sweet lady, and before long we all called her "Momma Anna." Their oldest child was a son named Masimo. I never got the feeling that Masimo liked us very much. That, or he just had a resting bitch-face. And then there was Mikala, their 18-year-old daughter. She was gorgeous, and I knew right away that the younger pilots would all be chasing her around like puppy dogs when they got here! After getting the keys to our rooms, I picked out the best one for myself and dropped off my bags before heading back to Aviano. Pulling away from the resort I once again noted the time and began the perilous trip down the mountain. It actually took longer by several minutes than the trip up because I had to ride the brakes most of the way down to avoid driving off a cliff!

When the squadron flew in the next day from South Carolina I met each four-ship of pilots at their planes. The Base Operations folks had kindly offered me one of their pickup trucks, complete with a combination UHF/VHF radio so I could talk to the jets as they taxied in. In preparation for the squadron's arrival, the other pilot who had arrived early with me helped me drive all of the cars over to the building we would use as our squadron. Afterwards we went to the Shopette on base and loaded a big cooler with ice and beer. As each pilot climbed out of their jet, we greeted them with an ice-cold brew and loaded them and their flight gear into the pickup truck. Dropping them at the squadron building, I handed each group of four a big envelop with their car and room keys, as well as map to Piancavallo. I told them where the cars were parked, but I didn't mention that they were sports cars, or that we were staying in a ski resort. By the time the whole squadron made it to dinner that night my prediction was right; I was a hero!

There was one unintentional thing that happened, caused because the maps to Piancavallo I provided showed the approximate driving time. All of the pilots began to compete to see who could make the fastest time in each direction. In just a few days the competition was heated. We even had a scoreboard in the squadron, and a list of rules. To begin with, all times would be measured between a certain stop sign at the foot of the mountain and the big billboard-like sign at the top saying, "Welcome to Piancavallo" in Italian. For a time to count, you had to have a witness who would initial the scoreboard in confirmation. Fighter pilots are a really competitive bunch, and the glory of a new "best time" was worth the risk of tumbling to a fiery death down the side of a cliff! The worst part of this competition was being a passenger in the car. I am sure that my body pumped more fear-induced adrenaline while riding as a passenger up and down the mountain than it ever did in combat.

You could actually go faster up the mountain because gravity helped you control the car around the hairpin turns. Generally, these uphill turns were made into the mountainside too. This added a little comfort for the passengers, because if you didn't make the turn you would crash into the side of the mountain and not go over a cliff. The downhill trips were the scariest, as guys would scream down the steep straight-a-ways as fast as the Alpha-Romeos would go, and then slam on the brakes to screech around the turns. When we would arrive at Aviano after racing down the mountain, the brakes would literally be smoking hot. Amazingly, we only had two crashes. The first was more funny than truly dangerous. The guy driving revved the engine and dropped it into first gear, peeling away from the road sign at the bottom of the mountain as his passenger started the stopwatch. After about a hundred yards there was the first 180-degree hairpin turn. The guy driving slowed for the turn, but didn't realize the road was covered in ice, and the car didn't turn at all. BAM! He drove straight through the turn and T-boned into the side of the mountain! Nobody was hurt, but the airbags did deploy, and the front of the car was crunched up. Not deterred, the four pilots piled out of the car and pushed it back onto the road. They discovered that it still ran fine, so they cut the airbags off with a handy pocketknife and continued racing up the mountain. This wreck resulted in a new rule: a crash that stopped the car would also stop the clock until you resumed but would result in a 2-minute penalty added to your time.

I was actually a passenger in the second crash, sitting in the right front seat. It was a downhill run. Passing the sign leaving Piancavallo there was a really long straight-a-way, followed by an ever so slight long curve to the left through a tunnel leading up to the first hairpin-turn. We flew down that first stretch, and the guy driving never let off the gas as we entered the curve in the tunnel. I could see right away that our rate of turn did not match the curve of the tunnel, and sure enough we ground against the tunnel wall. I sat there looking out the passenger side window at the wall of the tunnel literally inches away from my nose. A sense of dread washed over me as I realized the driver probably wouldn't be able to hold our line as the curve of the road tightened leaving the tunnel. Suddenly, the loud scrapping of the side panels on the tunnel wall was augmented by the smell of burning rubber from the tires rubbing against the concrete wall. All I could think of was, "Wow, I'm glad there is wall here to keep us from going over a cliff when the tires blow!" Soon we came racing out the other side of the tunnel. Fortunately, our grinding along the tunnel wall slowed the car enough that the driver was able to negotiate the ensuing curve, and we didn't go over the cliff. Our fearless driver didn't slow down a bit, and soon I stopped the clock as we crossed the finish line at the bottom of the mountain. It was a new downhill record, but we were all silently relieved when we parked at the squadron and climbed out. The side of the car was a mess, with torn fiberglass everywhere, a missing quarter panel, and the two tires that looked like suede leather. To add to the effect, smoke was boiling from the brakes too. All in all, it looked like a scene from a Mad Max movie after a big chase. To a man, I am sure we all felt safer flying our combat mission that night than we did commuting to work!

You would think that since every one of us was a terrified passenger several times a week that we would have stopped the contest, or at the very least have become sympathetic drivers. But we didn't. By the end of the deployment our fastest times had shaved almost 5 minutes off the trip each way, and I am proud to say that I was the record holder for the shortest uphill trip...an accolade I gladly share with my frightened passengers whose terrified timekeeping made it official!

Racing up and down the hill wasn't our only source of entertainment. We didn't have a ton of free time during the war, but we made sure that each pilot had one day off a week. You needed that time to do your laundry and decompress a little bit. During time off we made a lot of day trips in an effort to enjoy all that Italy had to offer. Aviano is in the far north of Italy, so traveling to Rome, Florence, Capri, or a lot of the big attractions and getting back the same day wasn't practical. Venice was pretty close, however, and that was a popular excursion. I really enjoyed Venice, even if the canals smelled, and went there several times. There was a beach nearby too, and when we realized that nudity was common on European beaches there was a mad rush to take in the sights. Of course, nude beaches are one of those things where the reality of the situation doesn't live up to your expectations. Yes, there were some beautiful topless women at the beach, but sadly it seemed like every one of them was accompanied by her 90-year-old topless grandmother. No kidding, one day when we were laying on the beach, a group from the local old folks home set up shop next to us. It was a less than glamourous experience, surrounded by sand, sun, surf, and friendly topless grannies. Even worse were the chatty grandpas in European-style banana-slings. Next to the base was the town of Pordenone, which was a pretty nice place to visit for some shopping. The single guys really enjoyed the nightclubs there too, and it wasn't uncommon to see a gorgeous Italian woman make the early morning walk of shame (or pride?) from the ski resort after coming home with one of our pilots. Piancavallo was a pretty remote place, especially after the ski slopes closed for the season. I always felt bad for these ladies when they had to go to the front desk in the lobby and ask Rocky or Anna to call a cab for them.

The debauchery wasn't limited to the nightclubs. There were only two businesses along the mountain road between Aviano and Piancavallo. One was a restaurant that specialized in serving wild game, but it was expensive so none of the pilots went there. The second business was a place called The Happy Boy, and it was a very popular destination. In fact, given the convenience of its location, guys would often change into civilian clothes before they left the squadron, and visit The Happy Boy on their way back to the ski resort after flying a mission. It was such a popular destination because The Happy Boy was a strip club. It was a pretty wild place, too, even by European strip club standards...but nowhere on the level of similar establishments in the Philippines or Thailand. It was always packed with gorgeous women who would do some outrageous things for the princely sum of 1,000 Italian Lira, which translated into less than one U.S. dollar at the time. Of course, I only knew of what happened at The Happy Boy from second-hand accounts; I always made sure I drove straight past to get home early so I could go to church in the morning. At least that's the story I'm sticking with.

Chapter 14: One Last International Incident

The Kosovo Air War ended on the 10th of June 1999. For a couple of weeks, we still continued to fly presence missions over Kosovo, but the cease fire meant that we wouldn't shoot unless attacked first. The Russians had also gotten involved, landing several transport planes of troops in Pristina to act a "neutral" peacekeeping force (although we all knew they were enthusiastic supporters of the Serbs). As part of their "peacekeeping" they also rolled out an SA-10 at the Pristina airport and kept it up and running. An advanced SAM like the SA-10 pretty much gave the Russians control over the airspace around Pristina and a good chunk of Kosovo. We could have dealt with it, if need be, but fighting an SA-10 would have been a whole different game. My generation of fighter pilots had started flying in the 1980s, and my Cold War experiences made me suspicious of the Russians, even if they were now our "friends." In fact, I still have a couple of old flight suits in my closet with "Fuck Communism" embroidered inside the collar!

I was a little surprised that the war was over so quickly, 78 days I believe in total. When we first arrived in Aviano at the start of the war, we were told that as the only state-side fighter squadron to deploy we would be there for the duration, and that was expected to be about a year. I wasn't concerned with that, and in fact I was excited at the prospect. I love flying combat missions. Nothing compares to the comradery, and the personal and professional challenges it presents. For me it was immensely satisfying as a man and as a fighter pilot to be good at war. And if you are an adrenaline junkie, which I am, the fix you get from combat is without comparison. I enjoyed flying in combat so much that after I retired from the Air Force, I found a job as a private military contractor (that's what we called ourselves, seeing as the term "mercenary" is now politically incorrect) flying high-risk security missions in some of the most dangerous places on earth.

At the end of June, it was time for the 78th Fighter Squadron to return to the U.S. and leave the peacekeeping mission to NATO's European members. For the flight home, I was scheduled to lead a six-ship, callsign Feud 31. The plan was to take off out of Aviano and meet up with some tankers who would take us across Europe and part of the Atlantic Ocean to the Portuguese Azores, where we would land and spend the night at Lajes Air Base. The next day we would again rendezvous with some tankers and fly the rest of the way across the Atlantic Ocean and back to our home base in South Carolina. The members of my formation were all highly experienced fighter pilots, except the guy flying as number six, who was still a young Lieutenant.

When the day came to leave, the six of us met up in the lobby of our hotel. After a quick cappuccino, we said our goodbyes to Rocky, Momma Anna, Masimo, and Mikayla, and piled into two cars for our last ride down the mountain to Aviano Air Base. Just outside of the base, there were always a huge number of anti-war protesters standing around, chanting, and waving signs. They had even set up a little "cemetery" of white wooden crosses to commemorate all of the "innocent" Yugoslavian women and children we had killed. As we pulled up to the main gate at Aviano, our young Lieutenant yelled, "Stop the car, stop the car!" We watched in amusement as he jumped out, dashed across the road, scooped up an armful of the white crosses, and dashed back to the car with a howling mob at his heels. Diving back in with his souvenirs, we peeled out, tires squealing, barely escaping the pissed-off protestors!

Soon we were in the squadron and huddled together for a flight brief before heading out to our jets. In the briefing I covered all of the things we would need to do on the flight. Even though this wasn't a combat sortie, flying half-way back to South Carolina was fraught with peril. In a single engine airplane with very limited gas, you can get into a lot of trouble if you have an engine problem, other emergencies, or can't get gas from the tanker. Another thing that was different was the route we would fly as we climbed away from the base. For the past three months we had been flying a departure routing that required us to observe noise abatement procedures, which meant that we had to stop using the afterburner as soon as it was safe to do so on takeoff. The afterburner was really loud, so it was important to minimize the noise effect on the town of Pordenone next to the base. For this flight we would be taking off out of Aviano on Runway 05, which meant the takeoff heading would be 050 degrees on the compass. The published departure procedure required us to remain within four nautical miles of the runway after takeoff, turning right to a heading of 270 degrees while climbing to 12,000 feet.

I was surprised that it said nothing about applying noise abatement procedures. I brought this up to my flight mates as I briefed, and everyone flipped through the other published departure procedures from Runway 05. Sure enough, all of the other departure procedures noted that noise abatement procedures applied. We had a good laugh about that, and that little moment of memorable levity in the briefing was actually what kept me from losing my wings and maybe getting Court Marshalled when we got back to Shaw AFB in South Carolina.

After the briefing we grabbed our gear and headed for the jets. There isn't a lot of room in an F-16, so when you are packing to fly halfway around the world you really can't take much with you. I would stuff my shaving kit into the little bubble of canopy behind the back of the ejection seat. I would carry a spare t-shirt, socks, and underwear in my helmet bag. When you added enough water and a little food for the eight or nine hour flight, your cockpit was pretty full. One of the guys in the flight decided he was going to take a case of wine in the cockpit too. It was apparently expensive wine, so he didn't want to pack it in his luggage and risk breaking it or getting it too warm and ruining it. We all watched in amazement when, after getting strapped in, he had the crew chief help him set the wooden crate with 12 bottles of wine in his lap! That was crazy! First of all, it would block him from seeing any of his flight instruments. At that time the Head's-Up Display in the F-16 was not certified for use as a primary reference of aircraft attitude, altitude, or airspeed when flying in the clouds. If we ended up flying in bad weather, he could end up losing control of the jet if something went wrong. Second, if he had to eject it was almost a certainty that the wooden case containing the wine would injure him badly. And last, and perhaps most problematic, there was no way he would be able to take a leak with that in his lap! When you fly for a long time, you will inevitably have to pee. Male fighter pilots do this using a Piddle Pack. A Piddle Pack is a long bag made out of thick, flexible, clear plastic with a funnel opening at one end, and a super absorbent sponge inside. After you fill it up with pee, the sponge soaks up most of it, and you use a twist-tie to close it up. It's pretty common after a flight of even short duration to see a pilot toss a full Piddle Pack or two in a trash can on the way back into the squadron. But with a crate of wine in your lap there was no way you would be able to use a Piddle Pack. As the crew chief climbed down the boarding ladder after delivering the crate, I smiled at the thought of how much fun we would have, giving this guy shit if he ended up peeing his flight-suit as a result of carrying the wine!

You may have noticed that a few sentences ago I specified that male fighter pilots use Piddle Packs to pee when they need to in the confines of a fighter plane cockpit. But what about the ladies? How do female pilots handle that? At first there wasn't a female friendly way to take care of that business, and I honestly don't know how the lady fighter pilots made do. Eventually the Air Force started providing diapers for them to use, but that was certainly a less than desirable solution. Fast forward a number of years, to when I was the Operations Officer of the 55th Fighter Squadron at Shaw AFB. The Air Force decided to finally address the peeing inequity for female fighter pilots head-on, and asked industry to come up with possible solutions. There were quite a few ideas and devices submitted to solve the problem. Some of them were quite painful, like having the woman catharize herself inflight and peeing through a tube into a bag. Another was an ultra-super-incredibly absorbent sponge that the female pilot was supposed to simply pee on. Some involved suction devices...an idea raised a lot of eyebrows and generated quite a few jokes. Anyway, the Air Force decided to test these different devices in flight, and they selected one of the female Flight Surgeons at Shaw AFB to conduct the evaluation. I just so happened to be pretty good friends with this particular Flight Doc, so she asked that I be her pilot for the test flight. I guess she figured it would be less embarrassing to fly with someone she knew while she tried out all of the different options. So, one fine day she climbed into the back of a two-seat F-16 with a bag full of items to test, and several big bottles of water. The crew chief for the jet didn't look so happy as he followed me around the jet during my preflight. He kept looking up at the Flight Doc in the rear cockpit, and I guess he was imagining having to clean up a pee-soaked seat cushion and ejection seat after we landed. My job was to simply fly around and give her the chance to pee a lot. The F-16 is equipped with the ability to record not only all of the displays and sensors in flight, but also anything that the pilot might say out loud into his oxygen mask. This is a really useful feature since you can just say what you want to remember later instead of pulling out a pencil and taking notes in the middle of a dogfight or flying a low level. It was also a great way for the Flight Doc to record her notes for her after-action report. We took off, and for the next four hours I flew up and down off the coast of South Carolina while she kept up a constant running dialog of her tests. We even had a tanker to make sure we had plenty of gas to test all of the options. She had been chugging water since before takeoff, so she was ready to get to work as soon as I sucked the landing gear up. While there was no way to see what she was up to in the back seat, for the next four hours I laughed until I cried listening to her describe what she was doing, step by step, squirt by squirt, the successes and the failures. She was a true medical professional engaged in her research, with a running dialog that was like listening to a combination science lecturer and football commentator. If I hadn't been able to use the autopilot, there were times when I might have crashed the jet, I was laughing so hard! By the time we landed I was dehydrated from the tears, and my ribs hurt like a football team had used me for a tackling dummy. As we walked back to the squadron after the flight, I told her I was going to write her up for a medal. She stopped walking and looked at me dead serious. "We will never speak of this again, and I had better not find out you shared the audio tapes with anyone." I smiled as I reassured her, "Don't worry, I will never tell a soul!"

Now, back to flying home from Aviano. After starting up we pulled out of the chocks and began taxing our six-ship out to the end of the runway for takeoff. As we rolled along all of our crew chiefs were standing along the taxiways cheering in celebration of a war well-fought. That set me to thinking, "What can I do to show our support for them, and give their morale a boost as well?" As we prepared for takeoff, I reviewed the departure instructions once more, and that was when the idea hit me. There wasn't a noise abatement procedure required for this departure, and the directions called for us to remain within four nautical miles of the runway while making a 220-degree right hand turn to 12,000 feet. The departure was designed to be flown in bad weather, so it didn't require any hard maneuvering. But today was a clear and beautiful Sunday morning, and since I was in an F-16 and had no noise restrictions, I could make this a pretty spectacular takeoff and departure; for sure it would motivate the guys on the ground!

I didn't tell the other five members of my flight what I had planned. I knew they would figure it out. Cleared for takeoff, we lined up in formation on the runway, and I gave the visual signal to run-up engines. With my engine stabilized and ready, I looked over and watched with pride as each wingman gave a big head-nod in succession indicating they were ready to go. I gave Feud 32 a salute, released the brakes, and pushed the throttle into full afterburner. My jet gathered speed in a hurry, and after I rotated away from the ground, I retracted the landing gear. Confirming that the gear was up, I leveled off about 15 feet above the runway and held it there to gain as much speed as possible. At the departure end of the runway, I was doing 450 knots and accelerating fast. I looked over my right shoulder to make sure there wasn't any traffic to worry about hitting, and then rolled into only a few degrees of right bank before pulling the stick back hard. I made the first part of the climb at about 6-Gs with the afterburner cooking, easing off the G once I was going pretty much straight up. In a second or two I added more G back in, and with a flick of the ailerons I rolled out wings level directly over the runway on a heading of 270 degrees at 12,000 feet...I followed the restrictions in the departure procedures to the letter! I pulled the throttle out of afterburner and watched as my wingmen followed me, each rolling down the runway at 30 second intervals. To follow me each one had to match my maneuver pretty much. Before long they had all rejoined on me, and we were on our merry way to the Azores. My undeniable spectacular departure had our people on the ground literally cheering. It also had the Italians screaming...for my blood!

As we were winging our way westward, I was blissfully unaware of a series of events unfolding at the highest levels of the Italian and American governments. The incredible noise from our afterburners happened to interrupt all of the church services going on that Sunday morning in the town of Pordenone, which is located right next to Aviano Air Base. The good people of Pordenone had been complaining about the noise of the jets conducting 24-hour a day flight-operations for the last several months, but the Italian government tolerated the complaints as part of their contribution to the NATO war effort. Now that the war was over, however, that patience was at an end. Within minutes of our takeoff the mayor of Pordenone was on the phone to the Italian version of the State Department. The Italians then pulled the U.S. Ambassador out of church in Rome to vent their anger, and even formally threatened NATO that Italy would prohibit flights from Italian air bases in protest. Only a year or so earlier a low-flying U.S. Marine Corps EA-6B had hit the cables supporting a cable car in the nearby town of Cavalese, killing 20 people, and the Italians saw this as another case of out-of-control American cowboy pilots goofing off at Italian expense. The political firestorm from our takeoff began to spiral out of control from there.

About the same time that I was leading our flight across Spain and Portugal, the U.S. Ambassador was on the phone with the U.S. Secretary of State, Madeline Albright, to alert her of what was going on. By the time we were coasting out of Portugal, Secretary Albright was calling her pal, Secretary of Defense Cohen, who in turn called the Chief of Staff of the U.S. Air Force and demanded that he ground the pilot responsible and launch a formal investigation. All of this happened unbeknownst to me, and happy as a clam I landed in the Azores. After the long flight we checked into billeting and joined the rest of the squadron pilots for dinner. Early the next morning we were airborne again to continue our journey across the Atlantic Ocean to Shaw AFB, South Carolina.

After an uneventful ocean crossing, we finally landed at Shaw AFB in South Carolina. There was a huge crowd of family, base personnel, and the press waiting to greet the conquering heroes. As we taxied in, I wasn't too surprised to see a group of Generals and Colonels waiting by my parking spot; I figured that since I was at one end of the parking row, they would shake my hand and move on down the line to the other jets in my flight of six. My family was there, and as I raised the canopy and climbed down the boarding ladder my nine-year old daughter, Katie, ran up and I scooped her in my arms for a big hug! As soon as I put her down, General Hawley, the Commander of Air Combat Command, walked up with an entourage. I assumed that he was there to shake my hand, and maybe even present me with his personal challenge coin for a job well done!

Boy, was I wrong. With little fanfare General Hawley informed me that by order of the Secretary of Defense, William Cohen, I was grounded pending the outcome of an investigation into my impromptu airshow departing Aviano. One of the Colonels with him wanted to interview me right that second to hear what I had to say. The General proceeded down the line to each of my flight mates, and while they were not grounded, each was whisked away immediately to give individual statements on the record about what occurred. I was pretty pissed off, but of course I cooperated fully…after all, cover-ups always get you in worse trouble than whatever crime you committed in the first place.

In the interview, the investigating Colonel listened to my story and took copious notes. When I mentioned that I fully complied with the published departure procedures, and that we had specifically noted the lack of noise abatement procedures for our assigned departure, he perked up. I reached down into my helmet bag and pulled out the official Department of Defense Flight Information Publication (DoD FLIP) that contained the departure in question. He read through it and compared it to the other departure procedures from Aviano just like we had done in the flight briefing back in Aviano before left. He stood up and said he would be right back. Apparently, he went out and called the other investigators who were interviewing the rest of Feud 31 flight to have them corroborate my story. Each pilot was being interviewed in isolation from the others so that the investigators could compare our stories. And of course, when asked, everyone told the same (truthful) story of our discussion during our flight briefing, noting that our departure was the only one not requiring noise abatement procedures. When all of the statements were finished, we were allowed to go. The Colonel kept the DoD FLIP that I had given to him, to reference later. I had 30 days of leave scheduled, so I wasn't too concerned about being grounded for the time being, but as I walked out of the squadron, I felt like the whole thing had put a damper on what should have been a banner day.

The investigation was quickly concluded. In the end nothing happened to me, and I was returned to flying status before I even returned from vacation. After the notes from all of our interviews were compared, the Colonels conducting the investigation were unanimous in their findings. The investigation revealed that the DoD FLIP departure we used that day had been printed with the noise abatement procedures inadvertently omitted. I had properly briefed the departure, and all of the flight members had corroborated each other's stories. Leading my flight on takeoff I had followed the departure procedures to the letter of the official DoD FLIP. It was an unfortunate incident, and perhaps my flying was too aggressive, but there was no reason to truly punish Captain Littleton.

In the end, Italy calmed down and did not restrict NATO flight operations. After returning from leave I returned to flying with no interruptions, and the whole thing quickly faded from everyone's collective memory as we got back to our normal peacetime training routine. But I will always look back at that day and proudly reflect that I am probably the only Air Force pilot in history to have been grounded by direct order of the Secretary of Defense!

Chapter 15: Airplane Crashes and UFOs

There are two questions that I get asked a lot: have you ever ejected from an F-16, and have you ever seen a UFO? I am happy to answer that first question by saying that the number of takeoffs in my logbook matches the number of landings, but I have had some really close calls. And a lot of my friends did indeed punch out of the F-16. Some more than once. The answer to the second question is yes, I have seen UFOs. Let me expand on both of those topics.

Flying high-performance jet fighters is dangerous business. On any given sortie, be it peacetime or combat, you are confronted with scores of opportunities to splatter your warm soft body across the ground. One of the things that makes a person able to fly fighters is having a supreme confidence in yourself. Literally, every day that you climb into a jet, you will bet your life on your flying skills and the judgement with which you use them. And when you are flying in a single-seat fighter there isn't anyone who can fly the airplane for you when something goes wrong.

Structural Failure

My first really serious emergency in the F-16 happened on takeoff one day from Prince Sultan Airbase in the southern Arabian Desert, during an Operation Southern Watch mission. The night before I flew the jet, it had undergone some significant maintenance that required the crew chiefs to unfasten some of the structural components at the back of the jet. When they put everything back together, they forgot to re-latch a series of fasteners around the tail that held on something called "the donut ring." The donut ring is a circular cowling that wraps around the back of the fuselage, just forward of the engine exhaust nozzle. Back then, when the fasteners are unlatched, they retracted inward into the fuselage, and were very hard to see. Checking these fasteners wasn't part of a normal preflight, so when I did my walkaround that day I didn't notice them. After my accident those fasteners ended up being re-engineered so that they would stick out from the skin of the jet when unfastened so that you can easily see them. So, blissfully ignorant of the problem, I climbed into the jet and headed off for a mission over Iraq. My callsign that day was Bill 22.

The jet was very heavily loaded, with 510 rounds of 20mm ammo, an ALQ-184 ECM pod, two external 370-gallon fuel tanks, two HARMS, two AMRAAMs, and two Sidewinder heat-seeking missiles. It was really hot, too, which didn't help the jet's performance. Regardless, the enormous amount of thrust produced by my GE-129 engine had me airborne in no time, and I began a left climbing turn to rejoin into close formation with my flight lead, Bill 21. Fighter pilots pride themselves on rejoining quickly and efficiently after takeoff, and I was concentrating on sticking the rejoin. I didn't notice anything odd about the way my jet was flying as I slid into tight formation on Bill 21's left wing. The guy who was Bill 21 had a reputation for being hard on his wingmen about little things, so I was determined to fly so perfectly he wouldn't be able to critique me later. A few seconds after I rejoined on him, Bill 21 started looking my jet over to make sure that during takeoff I hadn't lost any panels, had a hatch pop open, a gear door not shut all the way, or that I was perhaps leaking fluids. I was doing the same thing to his jet too.

Suddenly, Bill 21 started flying very roughly, his jet moving unexpectedly and abruptly. As I struggled to stay in tight formation, I thought to myself, "If he's going to fly like that, he better not say there was anything wrong with my formation flying when we get back from the mission!" We had already switched our radios from the control tower frequency to departure control, so I was surprised when Bill 21 told me to switch back over to the tower frequency. "Bill 21 check" he called over the radio. I crisply replied "Two," acknowledging that I had heard him. Still staring at my jet, and bobbing his airplane all around, he called out, "Tower, Bill 22 is an emergency aircraft and is going to High Key." High Key is a spot 10,000 feet directly over the center of the runway where you go in an F-16 if you think you are going to lose your engine. It took a second for it to sink in that he said Bill 22 and not Bill 21…he was talking about me!

Over our inter-flight radio Bill 21 called me up. "Bill 22, your engine cowling has come lose and you could lose your engine at any second." As he moved back away from me into a position where he could follow me at a safe distance and still help me out, I looked around the cockpit and at my instruments in dismay; I didn't see any indications that anything was wrong. All of my engine gauges were normal, and there were no Caution or Warning lights illuminated anywhere. When I thought of an engine cowling, I thought he must be talking about something on the front of the engine, but I wasn't sure. The pilot in an F-

16 sits on top of the engine intake, so there was no way for me to visually look under the jet to confirm my suspicions. As I turned toward High-Key I left the throttle alone. Most engine failures happen when a condition inside the engine changes, so not understanding what was wrong with my airplane I decided to leave the engine at its current power setting until reaching High-Key. Things were happening quickly, and as I was approaching High Key, I wanted very badly to understand what was wrong with my jet...I was at a loss because everything seemed normal. "Bill 21, describe exactly what you see wrong with my jet" I asked over the radio. Bill 21 replied, "The circular fairing around the back of the engine has come lose, and the slipstream has moved it back over your engine nozzle and is twisting things up." I shifted around so that I could look at the tail of my jet, and I instantly saw what he was talking about. The normally smooth and blended shape of the backend of the fuselage was all askew. I could see the yellowish chromate color with which the inside of the jet is painted. The top of the circular panel (the donut fairing) had caught the slipstream and had been pulled up and back. Not only was this bad because it could damage the nozzle which helped regulate the thrust of the engine, but it also threatened the engine's oil supply. The nozzle position on a

GE-129 engine is controlled by oil pressure, and the way that cowling was bent by the wind, it would be crimping those oil lines. If they cracked or snapped, the engine would fail in seconds as the oil squirted out. Suddenly it was all clear to me.

Fortunately, I was already at High Key by the time I figured out what was wrong with my jet. As long as I did my part and flew the jet right, it was aerodynamically possible for me to glide to a safe landing if the engine stopped working. At that moment it dawned on me that I had no idea if the jet would fly right at slower speeds with that cowling sticking up and out. I didn't want to find out for the first time if the jet was uncontrollable at low-speed close to the ground...that was something I wanted to discover while I had enough altitude to eject safely. So, I decided to stay at 10,000 feet for a little while longer to conduct a controllability check. The idea behind a controllability check is to configure the airplane for landing and slow down to approach speed while you are at high altitude. That way if something goes wrong you have some altitude to fix it.

I let Bill 21 know what I was up to, and level at 10,000 feet I flew away from the airfield but in a direction that would preserve my ability to make a dead-stick landing if the engine quit. I was going about 350 knots, and the maximum speed for the F-16's landing gear is 300 knots, so I took a deep breath, and slowly pulled the power back a bit. The engine continued to run fine, so I cracked it back some more, and passing below 300 knots I lowered the landing gear handle. The gear dropped down and locked into place normally. Even better, the jet behaved normally. I started slowing down some more to see if things stayed okay. If I could get down to 180 knots, which was about my approach speed with a full load of gas and weapons, I would know that I could make a normal landing. The F-16 lands with the speed brakes opened a bit, so I deployed those too. As the speed approached 200 knots the jet began to bunt noticeably nose-down. I added power, and I regained what felt like normal pitch control. That told me everything I needed to know; on final approach I needed to add 20 knots to the speed at which the nose began tucking under. That bit of information really got my attention. I would have to fly the jet down to the runway at 220 knots, and slowly bleed off the speed until I touched down, but I would have to touchdown just above 200 knots or the jet would pitch forward into the ground. To complicate matters, at 200 knots the jet would fly in a pretty much level attitude, and if I wasn't really smooth on the stick, I could land on the nosewheel first. If that happened, the jet would tip up onto the nosewheel and wobble like a wheelbarrow. Wheelbarrowing down the runway at any speed, much less 200 knots, would almost certainly lead to a fiery crash. If I was too aggressive at making sure the nosewheel didn't touch first, the jet would have enough speed that it would quickly porpoise back into the air nose high. If that happened, the airspeed would drop immediately below 200 knots, the nose would pitch down, and I would die in a spectacular fireball. And to put some icing on this cake, the nosewheel tire was only rated to 205 knots before it could come apart. The main wheel tires were only good to 215 knots before they could fail as well. In aggregate, I had a very small window of airspeed and pitch angle to pull off this landing, and even then the tires might blow at that speed and introduce a whole different problem. If my engine quit at any point, I might be out of luck all together. As I rolled onto final, I pulled the straps on my parachute harness tight and adjusted the ejection seat handle between my legs to make sure that I could reach it easily.

Heading down final I kept the speed wired at 220 knots. Bill 21 was behind me in chase formation and reminded me that I could jettison my external stores if that would help. I thought about it for a second, but I realized it wouldn't help since my airspeed limitations were due to the cowling and not my aircraft weight. Refocusing on the task at hand, I could see all of the firetrucks lined up near the approach end of the runway, lights flashing. Honestly, they were a big distraction sitting there, but there wasn't anything I could do about that at the moment. At 220 knots the runway came up fast, and it was time to finesse that beast onto the ground. I crossed the end of the runway and ever so gently rounded the jet into a level attitude, carrying the power until I was stable about five feet above the ground. Slowly, I cracked the power back bit by bit, and held the jet off until I could just perceive that I was one or two degrees nose high, then I let it sink toward the runway. At this point I wasn't paying attention to the instruments at all; this was one of those "fly-by-feel" moments. I'm sure I let out a huge breath when the main gear touched down, followed a microsecond later by the nosewheel! I wasn't out of trouble yet, however, as I had to get the airplane stopped...but getting into a controlled three-point attitude on the runway was a huge victory!

Now I was barreling down the runway at a very high speed. Normally F-16s slow down by holding the nose off the runway in a technique called aerobraking, but at the speed I was going I was certain the jet would fly back up into the air with disastrous results. So, with the throttle at idle and the speed brakes fully opened, I decided to be patient and see how it went. Prince Sultan Airbase has a long runway, about 13,000 feet long, so I had a lot of room to work with. About 1,200 feet before the end there was also an arresting cable stretched across the runway, and if I was still going too fast when I got to it, I could drop my arrestor hook and stop like a jet on an aircraft carrier. About halfway down the runway the jet was slow enough that I felt safe aerobraking, and I absolutely felt a huge wave of relief wash over me as I made it down to taxi speed. I rolled clear of the runway just like normal, except for the half-dozen fire trucks clustered behind me!

As soon as I stopped, I was surrounded by those fire trucks. After they determined that I wasn't on fire, a crew chief chocked my wheels, placed the safety pins in my weapons, and signaled for me to shut the engine down. As the canopy raised up and I took off my sweat-soaked helmet, the immense dry heat of the Arabian desert hit me. The crew chief attached a boarding ladder and climbed up, taking my helmet bag for me. After double checking that everything was shut down properly, I unstrapped and climbed out of the jet. A big crowd of people, including my squadron commander had gathered, and after a multitude of handshakes and congratulations I walked around the jet to survey the damage. Nobody knew what had caused the donut faring to come lose at that moment, but everyone was amazed at the job I had done getting the airplane down without bending it more or destroying it. A month later I learned that I had won the Air Combat Command Flying Safety Award for saving the jet and its full load of weapons. Unfortunately, it wouldn't be the last time I won that award.

Rapid Decompression

On that same deployment to Saudi Arabia, I was flying one day when my flight lead had a pretty unusual in-flight emergency. It had a humorous outcome and led to an interesting adventure. We had just come off the tanker and were crossing back into Iraqi airspace to the west of Kuwait around 25,000 feet. There was a fairly thick layer of puffy clouds around 15,000 feet, and I had just fenced in when, out of the corner of my eye, I saw my flight lead roll upside down and dive straight toward the ground through a gap in the clouds. Since he didn't say anything over the radio, and I couldn't see anybody shooting at us, I thought to myself, "A-hah! He's just trying to see if I am paying attention!" Without hesitation I jammed the throttle forward, rolled over, and dove after him. To my surprise, he kept diving straight down, weaving through the clouds. Following him, I soon punched through the cloud deck and saw my flight lead pulling out below, now headed eastbound into Kuwait. While I was very curious as to what he was doing, I also knew that a wingman is expected to not say anything on the radio except what was tactically required. There is an old fighter pilot saying that as a wingman, the only things the flight lead wants to hear from you is "2", "Lead you're on fire," and "I will take the fat one." A few seconds later, my flight lead called that he had experienced a loss of cabin pressure, and that we were diverting to Al-Jaber Air Base in Kuwait.

His having experienced a loss of cabin pressure at 25,000 explained everything. When you have a rapid decompression like that, it is vital that you quickly descend below 10,000 feet for a couple of reasons. First, a rapid decompression can cause you to lose consciousness very quickly if your oxygen system fails to function properly. Second, by getting below 10,000 feet, the ambient air contains enough oxygen that you can breathe if the airplane's oxygen system has indeed malfunctioned. Fortunately, my flight lead's systems worked as advertised.

He didn't say anything on the radio at first because he was busy dropping to a safe altitude.

I was excited to get to land at Al-Jaber. For starters, I had never been on the ground in Kuwait, and it would be cool to check that country off my "must visit" list. I was also excited because at our base in Saudi Arabia, we didn't have a chow hall yet, and we had been eating MREs for weeks. Al-Jaber was a more established base and was reported to have very good food! Finally, one of our sister squadrons, the 69th Fighter Squadron "Werewolves", had a detachment there pulling air-air strip alert. This meant we would have some friendly faces to hang out with and show us around.

We landed at Al-Jaber without incident, taxied to parking over by the Werewolf Vipers, and shut down. When we climbed out of the jets, my flight lead took off his G-suit revealing a large wet spot on his flight suit right over the crotch. When his jet abruptly lost cabin pressure he has just pulled out a plastic bottle of water to take a drink, and it promptly exploded due to the rapid decompression, soaking his lap. Most of it was on his G-suit, but the section of his flight suit that was wet over his crotch made it look like he had peed himself. When the Werewolf pilots who came out to meet us saw this, they howled with laughter! Immediately they accused him having wet himself when he had the emergency! The dry and insanely hot desert air soon dried the wet spot, but we told that story for years…and it's still a favorite to this day!

While maintenance fixed the faulty canopy seal on his jet, our brothers from the 69th gave us a great tour of Al-Jaber. This was 1996, so Desert Storm was only five years in the past. Al-Jaber had been taken over by the Iraqis, and during Desert Storm the USAF turned the base into rubble. While the runway and several facilities had been repaired and returned to use, most of the base was still covered in burned and destroyed structures. Every hardened aircraft shelter had a big hole in the roof where a laser guided bomb had punched through. Almost every building on the base was a burned-out shell. The paths and sidewalks that we strolled along were all bordered by yellow tape strung between stanchions with signs warning of unexploded bombs. And that was no joke or overly cautious warning…we could see mines and cluster bomblets laying around everywhere. One of the stops on the walking tour was the building that used to house the Kuwaiti Air Force Academy. From behind the yellow tape along the sidewalk out front, you could see a half-dozen skeletons in the foyer and on the stairs inside. The outside of the building was peppered with shrapnel and bullet holes but was largely intact and had not been bombed. But apparently some Iraqis had made a last stand in there when the Coalition troops cleared the base, hence some of the skeletons. The Iraqis had also booby-trapped the building so thoroughly that the other skeletons we could see were from people who had tried to go into the Academy once the fighting was over. Years later I was in a squadron with a Kuwaiti exchange pilot, and he told me they were never able to safely clear the building, and eventually it was blown up and a new Air Force Academy built in its place.

The highlight of our stay was, as anticipated, the food. Sadly, my flight lead's jet was fixed in a few hours, and we only had time for the tour and lunch. Not only did they have good food, but on one table they had bowls and bowls of candy bars for dessert! Before we left to go fly back to Saudi Arabia, we filled our flight suit pockets with enough candy bars to give one to everybody in the squadron when we got home. After weeks of MREs, these treats were so well received you would have thought we had delivered manna from heaven!

<u>High-speed Ejection</u>

In 2006 I was the Operations Officer of the 55th Fighter Squadron at Shaw AFB in South Carolina. Being the Ops

O of a front-line fighter squadron may be the best job on the planet. The squadron commander is always tied up with meetings and having to deal with the budget, personnel issues, and all sorts of administrative stuff. The Ops O's job, on the other hand, is to run the flying operations and make sure the squadron is ready for war. You make sure that everything that supports that goal is integrated, and it really pulls together everything you have learned over the course of your flying career. One of those responsibilities is making sure that your pilots receive qualification upgrades when their skills and leadership are ready for more in-flight responsibility. When a new young pilot arrives, they are simply wingmen, meaning that someone else leads them around the sky and into battle. Once a young wingman gains some experience and demonstrates good judgement in the air, they will undertake a training program to make them a Two-Ship flight lead. Two-Ship flight leads can lead a wingman around by themselves or be part of a four-ship formation flying in the number three position. Similarly, if a pilot does well as a Two-Ship flight lead, they will eventually go through an upgrade program to become a Four-Ship flight lead. And this process continues, with the best Four-Ship flight leads being selected to become Instructor Pilots (IPs), and then experienced IPs continue to become Evaluator Pilots who give checkrides. There are also dozens of other upgrades a pilot can be selected for depending on a squadron's specific mission taskings.

One day I was flying as an IP giving a Two-Ship flight lead upgrade ride. It was a BFM ride…dogfighting. Our callsign was Dice 21, and I was flying as Dice 22 that day with the upgrading pilot leading. Every fighter pilot loves to fly BFM, it's the sport of kings! When you fly BFM the goal is to put yourself in a position to shoot down the other pilot with the gun. The gun in the F-16 is a 20mm gatling cannon, and if you were actually shooting, it fires 106 rounds per second. It's a devastating air-to-air weapon but requires the pilot to put his airplane in a single position in the sky in relation to the other fighter to score a kill. BFM teaches a pilot not only how to survive in a one-on-one fight and kill with the most basic of weapons, but it also teaches a pilot how to fly his jet to the edges of the performance envelope and place it in an exact attitude at a desired speed and altitude. This ability directly translates to every other mission you fly. When we schedule BFM sorties the jets are configured with no wing tanks or heavy ordnance so that there are no restrictions on the speed or Gs. That means you have an F-16 that can sustain 9-Gs as long as the pilot can stand it and can fly out to beyond Mach 2 as well. A "clean" F-16CJ is the ultimate hotrod! For our mission that day we were assigned a large chunk of sky off the coast of South Carolina called W-177, right by Myrtle Beach and just north of Charleston. W-177 is a big chunk of airspace and BFM doesn't take up too much room, so we were sharing it with another two-ship, callsign Pistol 11. Pistol 11 was an IP upgrade BFM sortie.

By this time, the F-16CJ had received some pretty cool new technologies, two of which were Link-16 and the Joint Helmet Mounted Cueing System (JHMCS). Link-16 is a data-link system, and without boring you with details of everything it can do, one of its most valuable features is overlaying the position of other airplanes sharing the Link on a display in your cockpit. The JHMCS is a display that mounts on your helmet and projects a lot of data directly in front of a pilot's eyes. The biggest advantage of JHMCS is that you can target a lot of your weapons by simply looking at what you want to shoot. Another feature of enormous value is the ability to see certain Link-16 data in the JHMCS display as well. One of my favorite uses of this feature was to designate another airplane in Link-16 that I wanted to keep track of, and have that information displayed in my JHMCS. This is often referred to as putting a "bug" on someone. In this case, since we would be sharing the W-177 airspace with Pistol 11, I needed to make sure our two flights didn't accidentally run into each other while we were both dogfighting. By "bugging" Pistol 11 I would always have a line displayed in my JHMCS pointing to Pistol 11's location. Whenever I wanted to check on where Pistol 11 was at, all I would have to do is turn my head and follow the green line pointing to him. When he would be in my field of view the line would terminate in an octagon-shaped box around his aircraft.

The weather that day was good, but the winds were pretty high over the ocean, and the wave-heights were also high. There are limits on how high the winds and waves can be, because landing in high wind and wave conditions in a parachute is really dangerous, and you can easily drown. There are also limits on the water temperature. When the water is below a certain temperature pilots have to wear a water-tight anti-exposure suit to minimize the effects of hypothermia in the event they eject. On this particular day the winds and waves were right at the allowable limits, and the water temperature was one degree above the anti-exposure suit threshold.

It was late in the afternoon when we rolled onto the runway and blasted off in a formation takeoff. My young upgrading pilot did a good job leading us out to W-177, and before long we were ready to start fighting. As noted earlier, I "bugged" Pistol 11 so I could easily keep track of him and make sure our flights stayed well-separated. As a further safety precaution, both of our flights were sharing a common radio frequency in case we needed to talk to each other, and we would do our dogfighting on separate inter-flight frequencies. We had just finished our first engagement, when I heard the IP flying as Pistol 12 make an urgent call over the shared radio frequency, "Pistol knock it off, knock it off...Pistol 1 RECOVER, RECOVER!" I could tell right away from Pistol 12's voice that something bad had happened. Following the locator line on my JHMCS, I turned my head to find Pistol 11. I got my eyes on Pistol 11 and my heart just about stopped. I could clearly see him diving straight down toward the ocean at very high speed. A half-second later I saw the flash of his ejection seat rocket motor firing as he punched-out, and an instant later his F-16 dove into the ocean!

Pistol 11's airplane was going so fast and straight that is hardly made any splash at all. It reminded me of the small ripple you see when Olympic high divers enter the water. Without much thought, I used another feature of the JHMCS, which allowed me to mark a position on the ground by looking at it. The ocean is featureless and having a set of accurate GPS coordinates to search for Pistol 11 would be vital.

Pistol 12 called me very quickly and reported that Pistol 11 had ejected. I let him know that I saw it and would start coordinating a search and rescue response. When something like this happens, it is very important for all of the airplanes in the area to know what their roles are in the recovery, and to not run into each other as they scramble to help the downed pilot. I knew that Pistol 12 was surging with adrenaline, so I helped him out by organizing the response. Rightfully, Pistol 12 designated himself as the on-scene commander for the rescue effort. His job was to keep eyes on the downed pilot and try to make radio contact with Pistol 11 who would have a handheld radio as part of his survival gear. I told Pistol 12 to stay below 10,000 feet, and that I would keep everyone else above that altitude. The next thing I did was call up Jacksonville Center, which is the FAA air traffic controller for W-177. To make sure my radio call was heard I switched my radio to the universal emergency frequency called "Guard". "Jacksonville Center, Mayday, Mayday, Mayday. This is Dice 22 on Guard. Pistol 11 is down in W-177." Jacksonville Center immediately acknowledged and gave me a discreet frequency to call them on for further coordination. One of the problems now was that there were too many radio frequencies to monitor, as the F-16 only has two radios. I wanted to stay on the frequency that Pistol 12 was using to conduct the rescue, so that tied up one radio. I also needed to talk to Jacksonville Center on the discreet frequency to coordinate assistance from the Coast Guard. Additionally, I was mindful that all of our F-16s were configured for BFM, so we didn't have external gas tanks, which meant that I would need to also arrange for more F-16s to join the rescue. To facilitate all of this, I called up Dice 21. "Dice 21, I need you to climb and stay above 30,000 feet. Conserve your gas and call back to Shaw AFB and find out if there are any other F-16s nearby, preferably ones with external tanks that can stay on-scene for a while. I will be off frequency with Jacksonville Center but will call you back." With Pistol 12 staying below 10,000 feet and Dice 21 staying above 30,000 feet I had 20,000 feet of free airspace to bring in other airplanes and keep them separated in altitude as we searched for the downed pilot. I quickly changed over to Jacksonville Center's discreet frequency and passed them the coordinates of the downed pilot. The air traffic controller could relay that information to the U.S. Coast Guard, who would launch a rescue helicopter. I also let them know that there was only one person onboard, and the radio frequency that Pistol 12 was using as the on-scene rescue commander. Jacksonville Center let me know that the Coast Guard helicopter was on the way and would be there in about 30 minutes.

While I was coordinating the rescue helicopter, Pistol 12 was busy trying to contact his downed wingman. He had tried to contact him over the radio, but he couldn't hear any response to the radio calls. While working the radio Pistol 12 circled the crash site. He had seen a good parachute after Pistol 11 ejected, but the chute was soon swallowed up by the ocean. Searching visually, he couldn't see any signs of the wreckage, the parachute, or the downed pilot. Meanwhile, circling high above, my flight-mate Dice 21 had notified the base about the downed pilot and located some more F-16s that had just taken off with plenty of gas. They were a flight of two, callsign Killer 11, and both of the pilots in that formation were IPs which meant that we would have a lot of experience in the air. I directed Dice 12 to contact Killer 11 and have him arrive on-scene and maintain between 20,000 and 30,000 feet. With me staying between 10,000 and 20,000 feet everyone would have a safe block of altitude to themselves. The last thing we needed was a midair collision and more pilots in the water.

Killer 11 arrived on station about 5 minutes later, and I briefed him on the situation.

By this time Pistol 12 was running out of gas and needed to go home. As he departed, I assumed the on-scene commander duties and passed-off coordination with Jacksonville Center and the enroute helicopter to Killer 11. As I dropped down lower to search for Pistol 11, I realized just how bad the wind and waves really were. 15-foot waves mean there is a 30-foot difference between the top of the wave and the trough at the bottom between the waves. The winds were just high enough that the tip-top of each wave was blown into a white frothy spray. If that wasn't bad enough, the sun was getting low on the horizon, and that created deep shadows between the waves. I shook my head, "No wonder Pistol 12 couldn't find anything."

I decided that my best chance of seeing anything was to drop down low enough that I could see along the wave troughs with the sun to my back. I found the best combination was at about 200 feet, so that was where I circled, continuing to make radio calls in the hope that Pistol 11 could hear me. After a few orbits I got lucky, and for a brief moment, I saw Pistol 11's life raft as it crested a wavetop. The life raft in an F-16 survival kit is a one-man affair, and it is a blackish color designed to blend in with the water so that the enemy will have a hard time seeing you. I can testify that the camouflage is excellent. The raft is also designed with a little teepee-like shroud the pilot can pull over him to help protect him from the elements; fortunately, the inside of this "spray-shield" is bright orange, by purpose, to aid in rescue when the time is right. What I saw cresting the wave was a small triangle of that bright orange, and I adjusted my low-level orbit to circle that spot. Despite my best efforts, I never saw the raft again, and that was troubling. The raft is tiny and takes a huge effort to climb into. Given that I only saw a small orange panel of the spray shield I was concerned that Pistol 11 might not have made it into the raft with the waves and wind so bad. Regardless, it was better than not seeing anything at all, because the raft is tethered to the pilot, so at least I felt like I knew the general location of Pistol 11. I also used my JHMCS to mark the new coordinates and pass them to everyone else using Link-16. The coordinates would be a good starting point for the search, but I also was aware that the ocean has currents, and those currents would slowly move Pistol 11 somewhere else.

After a few more minutes I heard the helicopter check in; he was about five minutes out. I was already critically low on gas and decided that now was a good time to hand off my on-scene commander responsibilities to Killer 11 so that I would be out of the way when the helicopter arrived. I had Killer 11 drop down and fly above and behind me, and once he was saddled-up I flew over the spot where I last saw Pistol 11 and dropped out several flares to give Killer 11 a visual location to continue the search. I made one last call in the blind to Pistol 11, telling him that if he could hear me, Killer 11 was now in charge and that a rescue helicopter was on the way. Flying out of the area to the west, I climbed up and assessed mine and Dice 21's fuel states. Because Dice 21 had been orbiting at a very high altitude, he had conserved a lot of gas, so I instructed him to stay on station as long as he could, and act as a radio relay for Killer who was flying low and had limited radio range as a result. Leveling off at 16,000 feet I realized that I would have just enough gas to make it back to Shaw AFB and wouldn't need to make an emergency divert into Myrtle Beach airport. The flight back to Shaw only took about 15 minutes, and before long I was parked in the chocks with the engine winding down. I opened the canopy, took off my helmet, and just sat in the cockpit for a bit. I was sure Pistol 11 was dead. He wasn't the first guy I had known who died in a plane crash, and I knew he wouldn't be the last, but I was still crushed to lose not only a squadron-mate, but also a friend. I didn't know it at the time, but Pistol 11 was alive and getting rescued as I sat there!

As soon as I left the scene the Coast Guard helicopter arrived. The first thing they did was drop a buoy in the water at the last place Pistol 11 was seen. The buoy had a beacon in it that would allow the Coast Guard to find the location later if need be. It also had the added benefit of drifting in the current just like the downed pilot, which was very helpful. Killer 11 was climbing to a higher altitude to get out of the helicopter's way, when all of a sudden Killer saw a ship racing onto the scene. It was the U.S.S. Klackring, a U.S. Navy destroyer!

Almost an hour earlier, the Klackring was cruising along in the water underneath W-177. I don't know the reason, but they had a motion picture camera crew onboard, so everything I am about to tell you was recorded. It was dinnertime, and the Captain had just sat down with a group of his officers to eat, when in walked the second in command of the ship. They had just picked up my Mayday call and realized that they were close to the coordinates I passed to Jacksonville Center. With no hesitation, the Captain ordered the ship to proceed at flank-speed to help.

The Klackring arrived and quickly established radio contact with the Coast Guard helicopter, which was almost out of gas and ready to fly back to Charleston. The Klackring had a clear helicopter deck and offered the Coast Guard helo the chance land on the ship and refuel so that the rescue could proceed with minimum delay. As they were sorting that out, the Captain ordered the Klackring into their man-overboard drill to help search for Pistol 11. Standing on the bridge, the Captain observed that everyone was looking out the same side of the ship toward where Killer and the helo had been searching. The Captain, being an old and salty sailor, realized this wasn't optimum, so he instructed one of the sailors with binoculars to look out the other side of the ship. No kidding, like 30 seconds later, that sailor shouted, "There he is!" He had found Pistol 11!

At the beginning of this book, I wrote that I would avoid tying actual names to events, so for the rest of this story I will call Pistol 11" Bob." The Coast Guard helicopter flew over and spotted Bob and dropped a rescue swimmer into the ocean. The rescue swimmer made his way to Bob, who had somehow climbed into his raft despite having suffered horrific injuries when he ejected. After assessing Bob's condition, the rescue swimmer made a hard call. Bob was so busted up that getting Bob into a basket and hoisting him into the helicopter would surely kill him. Something else would have to be figured out. The swimmer was hoisted back into the helicopter, which then landed on the Klackring to refuel.

Back on the Klackring, word quickly spread of Bob's predicament. A group of older senior enlisted sailors huddled together, and quickly came up with a plan that demonstrated the ingenuity of crusty old shellbacks with decades of experience at sea. They put a launch in the water, and loaded it with a sheet of plywood, a bunch of two-by-fours, and some rope. Speeding out through the rough seas, they slid into position next to Bob. A couple of them jumped in the water and slid the sheet of plywood under Bob's raft, and in an amazing display of impromptu engineering skills, they used the two-by-fours and rope to raise the raft onto the launch without killing Bob. The launch sped back to the ship, and soon Bob was being hoisted up the side of the Klackring on this plywood platform. In another stroke of luck, the Navy doctor on the Klackring was a trauma surgeon by trade. He soon had Bob in his infirmary below deck and set about stabilizing his condition. I will discuss Bob's condition in detail later, but for now suffice it to say that when the doctor had done everything he could, Bob was loaded onto the now refueled helicopter and flown directly to the trauma center in Charleston.

I had just walked into the squadron as Bob was being loaded on the helicopter. The mood in the squadron was somber, and while most people knew something bad had happened, they didn't know exactly what had occurred or who it had happened to. I skipped going to Life Support to take off my flying gear, and instead walked up to the operations desk where the squadron commander was on the phone. I stood there, wondering how I should break the news, when the commander yelled, "OUTSTANDING!" He covered the phone with one hand and stood up, announcing with a big smile "They found him, and he is alive!" A huge cheer went up, and I won't lie…my eyes teared-up.

The next morning, I drove down to Charleston to see how Bob was doing. He was a mess. But he was an alive mess, and it looked like he was going to live. Bob was still unconscious, which was probably for the best. His head was the size of a pumpkin, and his face was covered with bloody scraps. The rest of his body was pretty much surrounded by a cage of metal braces and wires holding his arms and legs together. I spoke to his wife for a while and let her know I would check back with her frequently, and to let me know if they needed anything at all. Both her and Bob's parents were on the way, but I wanted her to know the squadron was her family too and would give her and Bob any support that they needed.

Of course, there was an investigation, and here is what it revealed. At the start of their second dogfight, Bob was pulling 9-Gs as he maneuvered to get on Pistol 12's tail for a gun shot and passed out from lack of blood to his brain. This is G-induced Loss of Consciousness is called G-LOC for short. G-LOC incapacitates a pilot for anywhere from about 12 to 30 seconds before a pilot recovers enough to be able to fly the airplane again. This may not sound like much, but when you are in an F-16 in full afterburner you most likely won't have that long before you slam into the ground at high speed. G-LOC is a frequent killer of fighter pilots. As Bob passed out, his IP flying as Pistol 12 immediately recognized what had happened and had made the urgent radio calls I had heard for Bob to recover his jet. In the cockpit of his F-16, Bob found himself slowly regaining consciousness. He noticed that the ocean was rushing up at him, and in an amazingly heroic feat managed to grab the yellow handle between his legs and eject.

The investigation revealed that if Bob had pulled the handle a half-second later his ejection seat wouldn't have cleared the jet before it hit the water. When Bob ejected, he was still half unconscious, and his body posture wasn't right for an ejection. He was still holding onto the stick with his right hand when the seat fired, and the shock of that tore his shoulder out of the socket and broke his right arm into a bunch of pieces. Being slumped over, the force of the rocket in the ejection seat bent him over even further, which messed his back up really bad too. But Bob's problems were only getting started when the seat left the jet. Heading straight down in full afterburner, the jet was almost supersonic, and the blast of wind at that speed was like hitting a brick-wall. It did horrible things to Bob's body. The JHMCS is only approved for ejections up to 450 knots, and Bob was way faster than that. Bob's JHMCS was ripped off his head by the wind blast, taking a lot of flesh with it. And even worse, the wind folded his legs back and under the ejection seat. The force of the wind form-fitted his legs to the seat so violently, that in addition to breaking his legs, it actually ejected several inches of bone from his shins. Bob was in bad shape when he landed in the ocean. Fortunately, the seawater activated releases on his parachute worked, so the canopy didn't drag him under when it sunk into the ocean depths. Likewise, the life preservers attached to his harness activated as well, keeping him afloat. Bob didn't recall how he climbed into his small life raft, but he somehow managed. Fighter pilots have to regularly go through refresher training on our survival equipment, and that includes having to climb into the one-man life raft. It's always a source of amusement watching guys kick and struggle trying to climb into it, and it usually takes several tries in the docile conditions of a swimming pool to make it aboard. The fact that he managed to do it with his injuries and the high waves is an absolute miracle. Another miracle was the cold-water temperature. The surgeon on the Klackring said that the cold water put him into a near hypothermic state, which kept him from bleeding-out. Incidentally, Bob's first memory after regaining his senses in the life raft was seeing an F-16 flying over him dropping flares.

Bob was told by the doctors that he would live, but that he would never walk again. But the kind of man who can survive a high-speed ejection doesn't let anyone tell him what he can or can't do. About six months later the squadron threw a going away party for Bob and his wife. Bob stood up from his wheelchair to accept his farewell plaque. There wasn't a dry eye in the room! Bob eventually did learn to walk again, and then he began to run. He made such a miraculous recovery that he was able to pass a flight physical and return to flying status. Because of the injuries he survived, it wasn't safe for him to fly in an ejection seat aircraft anymore, so he went on to fly C-130s. Bob is one of my heroes, and I hope I get to buy him a beer one day and hear more about the second chance at life he and his family received!

As for me, Pistol 12, and Killer 11? We were awarded the Air Combat Command Safety Award for the search and rescue. A few months later the 55th Fighter Squadron Commander met with the Captain of the U.S.S. Klackring and presented him with a case of very fine whiskey to share with his crew in appreciation for their quick action that day.

<u>Chum in the Water</u>

I am happy to say that not all ejections end in remarkable stories of survival. In fact, some are pretty humorous, and that makes them fun stories to tell. Take for instance the day Fang 22 ejected.

Fang 21 and Fang 22 set out one beautiful summer day to go dogfight each other off the coast of South Carolina. Fang 22 was flying an F-16CJ with the tail number 91-397, which was the first F-16 that had my name painted on the canopy rail. Before starting their dogfights, Fang flight executed what is called a G-warmup exercise. A G-warmup exercise involves making a couple of hard turns, with the first one around 4-5 Gs, and the second one at 7 or more Gs. The point of this exercise is to make sure your G-suit is functioning and that your body is up to the high-G environment before you experience the prolonged 9-G environment of a dogfight. After the second turn of the exercise Fang 22's engine rolled back to idle, and his jet began a slow gliding descent. Having plenty of altitude, Fang 22 followed all of the checklists to solve the problem, but nothing worked...the engine simply wouldn't accelerate past idle RPM. When it became obvious that he wouldn't be able to get the engine to produce thrust, and he wouldn't be able to glide to an airport for an engine-out landing, Fang 22 prepared to eject. He wasn't too concerned, as it was a beautiful, warm summer day, there was nothing but a gentle breeze, and there were no waves at all. He was also pretty close to Myrtle Beach, so the rescue wouldn't have to take place far out at sea. There is a checklist you can run in the F-16 if you have the time before ejecting, and Fang 22 followed it to the letter. Fang 21 had already called Jacksonville Center, and the Coast Guard helicopter was on its way. At 2,000 feet above the water Fang 22 zoomed the jet nose high, and at its slowest point at the apex of the zoom, he pulled the yellow ejection seat handle between his legs and rocketed out of the jet.

Everything worked as advertised, and Fang 22 was soon floating comfortably under his parachute, gazing at Myrtle Beach in the distance with Fang 21 flying in circles above him.

All of this had unfolded so smoothly that Fang 21 was pretty much a spectator as his wingman splashed into the ocean. Fang 21 noted the position and passed the coordinates to the Coast Guard helicopter, which was only about 20 minutes away. Meanwhile, Fang 22 was bobbing around in the water trying to climb into his one-man life raft. Once aboard he assessed that he was physically okay and uninjured, so he decided to pull out his survival radio and call Fang 21. The funny thing is the radio was damaged somehow, and every time Fang 22 would key the mike to speak, he would get a small electrical shock! Despite the discomfort, he let Fang 21 know he was okay, and upon hearing that the rescue helicopter was on the way, sat back to enjoy his time in the sun.

Orbiting overhead, Fang 21 was starting to get boarded, so he decided to have a little fun. Over the radio he called Fang 22. "Fang 22, this is Fang 21. You might want to keep your arms and legs out of the water…there's a really big fish swimming around you." Fang 21 was implying that there was a shark, just to tease his buddy in the life raft. Fang 22, however, did not find this funny in the least, and sat there wondering when Jaws would decide to take a bite! Soon, the Coast Guard helicopter arrived and began to hover nearby.

Fang 22 said that it was an amazing sight to see the rescue swimmer drop from the helicopter. The guy was hugely muscular, had close-cropped blonde hair, and for all the world looked like Dolph Lundgren! Even more incredible, the rescue swimmer swam to his raft so fast it looked like he was skipping across the water instead of swimming in it. When the rescue swimmer got to the raft, he held onto the side with one hand and dropped his snorkel out of his mouth. He asked Fang 22 is he was injured, and even though Fang 22 said no, the swimmer proceeded to check his arms and legs to make sure there were no bones sticking out and that Fang 22 wasn't simply in shock and didn't notice them. Determining that the pilot was indeed okay, he instructed Fang 22 to climb out of the raft. "I'm not getting out of the raft," Fang 22 responded, "there's a big shark out here!" Unphased, the rescue swimmer replied, "Don't worry about him, the ocean is full of big sharks, he won't bother us. I need you to get out of the raft so we can swim back under the helicopter and get hoisted up. We can't fly the helicopter over your raft, because the rotor-wash might suck the raft up into the helicopter's blades and make it crash." But Fang 22 wasn't backing down, "I'm not getting out of the raft!" The rescue swimmer nodded, apparently resigned to the situation, floating there calmly with one hand still on the side of the raft. Suddenly, Fang 22 noticed that the raft was deflating and sinking under him. The rescue swimmer smiled at him and held up his other hand that had been under the water; he was holding the dive-knife he had just used to poke a hole in the raft! "I guess you have to get out of the raft now, Sir!"

Accepting his fate, Fang 22 rolled out of the sinking raft. The rescue swimmer told him to relax and let him tow the pilot back to the waiting helicopter. Fang 22 later told us that the rescue swimmer grabbed him by the back of the collar and started swimming so fast that Fang 22 felt like he was wake-boarding behind a speedboat! In short order Fang 22 was safely aboard the helicopter and was flown to a nearby hospital for a checkup before being returned to Shaw AFB. We had a great impromptu party when he walked back into the squadron that afternoon. Fang 22 and I were close friends, and I was very happy to see he was okay. But I was pissed that he crashed my jet!

<u>Rattlesnakes</u>

Another ejection story that ended well was the crash of Widow 41, when I was stationed at Luke AFB in Arizona. At the time, Luke AFB was where almost all of the new F-16 pilots went for their initial training in the jet. Luke is a great place to fly fighters. For something crazy, like 300 days of the year, the sky is perfectly clear. The weather is always warm, if maybe too hot for a few months during the summer. Away from the cities the desert is mostly devoid of human habitation, so there are a lot of huge bombing ranges to use. The mountainous terrain also makes for some terrific low-level flying. In addition to several squadrons that teach U.S. pilots how to fly the F-16, there are a couple of squadrons that teach foreign pilots to fly the F-16 for their home Air Force's. The only problem with the F-16s at Luke, was that they all had Pratt and Whitney engines. The F-16 can accommodate either GE or Pratt engines. I guess this was to support the defense industrial base by giving both engine manufacturers a slice of the pie. But the reality is, the GE engines are a lot better than the Pratt engines. The GE engines rarely failed, while the Pratt motors had a bad track record of quitting or flying apart. During the Kosovo Air War there was something like a two- or three-week span of time in which Luke lost five or six F-16s due to engine failures. Our joke was that we were safer dodging SAMs over Belgrade with a GE engine than we would be teaching a new pilot to land in a Pratt powered F-16 at Luke!

One day a U.S. instructor pilot was leading a foreign student pilot out to one of the bombing ranges in southern Arizona. Everything was going great as they cruised in line-abreast tactical formation, when suddenly the U.S. instructor pilot, callsign Widow 41, felt and heard a huge bang that literally threw him forward against his straps. At that same moment just about every Caution and Warning light in the cockpit lit up, and the aural warning system started voicing all kinds of bad things. Widow 41 turned around and looked back at his jet and saw a huge wave of flames boiling up and forward toward the cockpit. While F-16s are not prone to exploding, even when on fire, that was all he needed to see, and immediately punched himself out of the jet. The ejection went fine, and Widow 41 watched from the safety of his parachute as his flaming jet slammed into the desert floor. Soon he prepared for his own landing in the remote desert, and with a thump found himself laying on his back with the parachute draped over him. Widow 41 laid there for a second catching his breath and slowly moving his arms and legs a little to make sure nothing was broken. Feeling confident that he was okay,

Widow 41 sat up, still covered in the parachute. As he sat up and started to tug at the parachute, he was mortified to hear the rattle of a big rattlesnake somewhere under the parachute with him! Widow 41 instantly froze. The rattlesnake was somewhere right in front of him, not more than a few feet away. Very slowly he raised the parachute in front of him, there it was...a huge and very angry Western Diamondback, coiled up between his feet! Just as slowly, Widow 41 lowered the parachute back down. He was sitting there considering his options when the rattling stopped. At first this was a relief, but then Widow 41 realized the silence could mean the snake was trying to crawl away and might crawl right into him in the process! After sitting still for what seemed like forever, Widow 41 once again raised the parachute very slowly. He was relieved to see that the snake was gone and nowhere that he could see. Very slowly and carefully, he pulled the parachute over his head from back to front until it was in a bundle between his legs. With great caution he slid his legs back out from under the chute and finally jumped to his feet.

That crisis solved, he became aware of the sound of his student wingman, Widow 42, orbiting high above. Widow 41 pulled out his survival radio and quickly established radio contact with his wingman. Now he discovered that he had a new problem; the student was so shaken by his IP's jet crashing that he couldn't calm himself down enough to speak in English! Widow 41 tried to assure his wingman that he was okay in an effort to get things back under control. He needed Widow 42 to call for help and get the rescue mission started. But it was of no use, and his student continued to chatter nonstop in his native language. At that very moment Widow 41's cellphone rang.

Widow 41 eagerly dug his cellphone out of his pocket, expecting to see that it was someone from the squadron calling him. He was shocked to see that the call was from his wife! He answered it, and they chatted about normal things, as Widow 41 tried to figure out how to break it to his wife that he had just punched out of a jet and was standing in the middle of the Sonoran Desert nervously watching for rattlesnakes. As the call wound-down, he finally said something like, "Hey babe, I need you to do something for me. I need you to call the squadron and let them know I ejected on the way to the range and need them to send help." It's a real shame he couldn't have recorded the rest of that conversation because I bet it was a real doozy! I am quite sure she was horrified, but she apparently was much calmer than Widow 42, and after hanging up with her husband she called the squadron as he asked. Widow 41 was rescued with no further drama, and to this day I am willing to bet that his wife is the only spouse to have initiated the search and rescue for her downed airman!

Two pilots, One Parachute

Perhaps the most incredible, and unbelievable, story about an ejection that I know of occurred out at Luke AFB, in Arizona. Luke AFB was, at the time, the primary location for certain specialized F-16 training. One squadron, the 310th Fighter Squadron, taught courses on flying with Night Vision Goggles, employing targeting pods, and how to be a FAC-A, which stands for Forward Air Controller-Airborne. The FAC-A course was especially demanding, and as such, the students selected to attend were some of the most experienced F-16 pilots in the Air Force. All of the training sorties in the FAC-A course are flown in a two-seat "D-Model" F-16. The reason for this is that the instruction is quite intensive and is easier done over an intercom in the same airplane than over a radio from another jet. On a typical training mission, the student and the instructor fly as a single-ship and meet up with a series of fighters over the training ranges. The student then identifies targets and works the complex coordination required to mark those targets as hostile, and then sequence the fighters onto the target based on what ordnance they have and what will be most effective against that particular type of target. Some of the scenarios are flown at low altitude in a high-threat environment, with notional troops on the ground. Other scenarios are carried out from medium altitude with different ROE and coordination requirements. The FAC-A training missions were time consuming and difficult, but when you graduated from that course you knew your stuff. I had the chance to attend the FAC-A course and to be FAC-A in combat, and it was an enormously satisfying mission!

One day two of my old friends, Nutman and Tonka, where going out to fly a FAC-A upgrade sortie. Nutman was the student, and Tonka was riding in his backseat as the instructor. Everything was going great until they reached the end of the runway before takeoff. Their F-16D model developed some problem that they couldn't accept, so they taxied back to parking to get another jet. The timing of FAC-A training missions was critical, because you had several other fighters you needed to meet over the bombing ranges, and if you were too late, they would run out of gas before you could get your training done. So,

Nutman and Tonka were in a hurry when they climbed into the spare jet.

When you fly in a two-seat F-16 there are several things you have to check in the backseat that are different, depending on if you are flying with the backseat occupied or empty. Chief among these is the way the ejection seats will work. If there are two pilots in the airplane you have to set the system up so that when you pull the handles you get what is called a sequenced ejection. Basically, regardless of which pilot initiates the ejection, after the canopy leaves the jet, the backseat ejects first, and then the front seat fires. This happens so that the two seats don't hit each other. If you are flying a two-seat F-16 with the backseat empty, however, you don't want to delay the front seat ejection by the second or so it takes for the unoccupied backseat to go first. All of that makes perfect sense. But for a scenario that no one has yet provided me with a good explanation for, the two-seat F-16s have a third option for the ejection system. This odd setting allows each pilot to eject separately without any sequencing; you eject whenever you pull the handle, but the other pilot stays in the jet until he decides to pull the handle on his own. To select which type of sequenced or non-sequenced ejection you have available, there is a selector T-handle in the backseat. When you fly alone, you always check that handle is in the "Solo" position before your go fly. If you are flying in the backseat, one of the checklist items is to make sure you turn the handle to the position that allows for a sequenced ejection. I can't think of any scenario where I would select that third option. Nutman and Tonka were in a hurry when they hopped into the spare jet that day, and Tonka didn't notice that the ejection selector handle was in that third position.

On takeoff, shortly after retracting the landing gear, the engine in their F-16 started coming apart. Nutman jettisoned the external tanks and ordnance and began a climbing turn to gain enough altitude to land back at Luke AFB if the engine quit completely. Unfortunately, it did quit before they were in a position to attempt an engine-out landing. Being at low-altitude, and with very little airspeed left, Nutman did the right thing and zoomed the jet upwards before commanding a bailout.

But because the ejection mode handle in the backseat was in the wrong position, Nutman and Tonka did not get a sequenced ejection. Instead, Nutman in the front seat went up the rails just a fraction of a second before Tonka. At low altitude the ACES II ejection seat is smart enough to recognize the situation, and immediately kick you out of the ejection seat and deploy the parachute. This is exactly what happened with both ejection seats, but because they did not have a sequenced ejection, the rocket motor in Nutman's seat burned through the shroud lines of Tonka's chute as it began to deploy. As Nutman's chute deployed he slowed down immediately. But Tonka, now free of his seat but with no parachute, flew forward and smashed into Nutman. In an amazing strike of luck, Tonka found himself entangled in the shroud lines connecting Nutman to his parachute canopy! Because of the low altitude of the ejection, they basically got one swing of the parachute before they hit the ground. Neither one realized that they both came down in the same parachute until later, when the accident investigation was complete!

But their good luck didn't end there. It turns out they ejected over a farmer's field full of cows and farm equipment. The spot they landed in was in the middle of all sorts of tractors and plows...basically, if they had landed just a few yards in any direction they would have likely impaled themselves on sharp metal. Nutman and Tonka climbed out from under the parachute and were in the process of asking each other if they were okay when a herd of cows stampeded toward them. The cows had been grazing nearby when the jet crashed, and the explosion spooked the herd, which stampeded away from the burning jet…and straight toward Nutman and Tonka. Thinking fast, the two pilots dodged the leading cows and made safe refuge behind some of the farm equipment that had so recently threatened their lives! Pretty soon they were picked up by a helicopter and taken to a hospital. They were released with only minor injuries, but perhaps one of the most incredible ejection stories of all time: engine quit on takeoff, we ejected at low altitude came down in the same parachute, landed safely in the middle of a bunch of farm equipment, and then avoided being trampled by a stampeding herd of cows!

Out of Control

The F-16 supposedly can't be put into a spin, according to the engineers. But I can testify that is does a pretty good imitation of one if you put the jet into the wrong place aerodynamically. The key to understanding how an F-16 goes out of control is to understand that a computer controls the airplane. The F-16 was intentionally designed to be inherently unstable. This is the secret to the F-16's fantastic turn performance…it wants to pitch nose-up. A person can't react fast enough to keep the airplane under control due to its inherent instability, so a computer flies the airplane based on what the pilot asks it to do. The computer senses the pilot's control inputs, looks at the jet's angle of attack, airspeed, etc., and then commands the flight controls to do what the pilot wants. But there is a caveat to that; if the pilot asks for something that the computer thinks will put the airplane of out control, it won't do it. It will only give the pilot some of what he asks for up to the limits of aerodynamic control. This has lots of advantages, including that for any given combination of airspeed, altitude, angle of attack, etc., the computer will give the pilot the maximum performance available. The problem is, every now and then, a pilot gets the jet into a position that the computer can't handle. This usually occurs when the jet is slow with the nose high above the horizon. When that happens, the jet can depart controlled flight.

When an F-16 goes out of control, the first thing the pilot does is nothing. You let go of the stick and see if the jet will slip back into its normal aerodynamic envelope as gravity takes over. Most of the time the jet will run out of airspeed, and the nose will drop or slice sideways over toward the ground. As speed picks up, air flows properly over the wings, and the jet is controllable again. But sometimes the nose doesn't drop enough to restore airflow over the control surfaces, and the jet enters a condition known as a deep stall. Basically, the airplane stops flying and falls out of the sky in a flat attitude, or sometimes even slightly tail-first. If this happens the pilot can engage a switch call the Manual Pitch Override switch, or MPO switch for short. Engaging the MPO switch allows the pilot to have more control over the horizontal stabilizers without the computer getting in the way. With the MPO switch engaged, the pilot rocks the nose of the airplane up and down until it pops out of the deep-stall and starts flying again. Take my word for it, it's a lot trickier than it sounds! While in the deep stall, the airplane can spontaneously perform lots of different post-stall gyrations. It can yaw, it can roll, it can flip inverted…or do all three at the same time which makes for a violent ride. Combine all of that with the fact you are falling like a leaf to the ground, and it can be very disorienting. The F-16 community has lost a few jets over the years due to loss of control. Because it can happen, we practice recovering from out-of-control situations in the simulator, but never in the actual airplane. Most pilots will never put their F-16 completely out of control, but if they do, their first opportunity to recover from one will be for real! Interestingly, some highly experienced F-16 instructor pilots are selected to go out to Edwards AFB and put an F-16 out of control for real, with an experienced test pilot in the back seat of an F-16 configured with a spin-recovery chute. The idea of the course is to explore and experience the post-stall characteristics of the F-16. I was lucky enough to be selected for this course.

The course included just me and one other pilot and was scheduled for a Friday. The morning of the flight, we both went over to the USAF Test Pilot School, where we met up with the test pilot we would fly with. After shooting the shit for a while over a cup of coffee, the three of us settled into one of the classrooms and spent a few hours reviewing the flight control system of the F-16, diving deeply into the flight control logic and capabilities. We also worked through the aerodynamics associated with the edges of the F-16 flight envelope, and what happens when the F-16 crosses from controlled flight into uncontrolled flight. It was fascinating stuff, and very educational. After class, we three went and grabbed lunch, and then went to fly. The other pilot and I played "rock, paper, scissors" to determine who would fly first, and I won!

I had flown out of Edwards AFB many times before, but never in one of the white and orange Test Pilot School F-16s. The jet we had that day was a two-seat F-

16B model, manufactured in 1978, making it one of the first F-16s made and delivered to the Air Force. It was equipped with a small parachute on the back of the jet, called a spin-chute. While you can predictably recover an out-of-control F-16, it is very possible to get it into a situation where you can't return to controlled flight. If that happens, the spin-chute is deployed, and that will yank the plane into a tail-up/nose-down attitude, allowing air to flow back over the wings like normal. The spin-chute can be jettisoned once used. Our jet that day was also configured with a centerline gas tank. I was excited to see this, as a two-seat F-16 with a centerline tank is very unstable post-departure and promised to make my ride a wild one! I strapped into the front seat, and soon we were rolling down Edwards AFB's insanely long runway.

If you are a military pilot, Edwards AFB carries a special meaning because of all the famous pilots and flying that have occurred there. Literally every kind of airplane has been flown and tested there, the sound barrier was broken there, X-15s blasted into space from there, and Space Shuttles have even landed there. As we climbed up to 45,000 feet for the first departure, I looked down at the Edwards' paved runways, as well as the dry lakebed that has an enormous 25,000 foot long runway marked on it. I grinned like an idiot thinking, "How cool is it that I am about to put a jet out of control here like Chuck Yeager!" I did 13 departures in total, and every single one was the wild ride I hoped for. We did the departures from a high altitude so that we could let the jet stay out of control for a long time. As we fell toward the earth, the test pilot in the backseat would have me look outside and watch the flight control surfaces as the computer tried to recover the airplane, explaining what we saw just like he described during the morning's academics. Passing 25,000 feet he would conclude the academic lesson, and then I would recover the jet. All 13 times I was able to recover just fine, so the spin-chute never came into play. The flight only lasted a little more than an hour, but it was one of the more educational and just plain fun experiences I've ever had in an airplane! After a debrief and review of the video tapes from the flight, I caught a ride to LAX and was home in time to fall asleep in my own bed.

And wouldn't you know it, on Monday morning I put my jet out of control for real! I was scheduled to lead an ACM ride, callsign Medusa 11, with two bandits instead of just one. Our training scenario that day was what we called "Tap the CAP." Basically, me and my wingman would orbit at a predetermined point at 20,000 feet. We wouldn't use our radar, instead relying on visual lookout to spot the bandit. The two bandits would stay nearby, and one of them would try and sneak up on us unobserved. When we saw him, we would react and fight as a team to kill him quickly in a 2 v 1 dogfight. As soon as we would kill the bandit, we would maneuver to get back into a tactical formation so we could support each other visually and await the next attack. The two bandits would coordinate so that very soon after each kill, the other bandit would commence another attack. Tap the CAP was always fun, and a good pair of bandits tag-teaming you would make for some challenging fights.

My wingman and I had been fighting for quite some time when it happened. We had just killed one of the bandits, and we came out of the fight one behind the other. Medusa 12 was in front of me, so to regain visual mutual support, I had Medusa 12 turn 90 degrees one way so that we rolled out side by side about a mile apart. The very instant we rolled out, my RWR went off, alerting me that a bandit was somewhere above me, locking my jet with his fire control radar. Twisting around in my seat and looking up, I saw the bandit diving in from high behind us. It looked like the bandit was pointed at me, so I made a call over the radio, "Medusa 12 break right, bandit my high Six

O'clock!" As my wingman executed a high-G turn to his right, I pulled straight up into the bandit to cause him to overshoot me. Not only would this spoil his shot against me, but it would also force him to choose between fighting me or turning after Medusa 12. This was perfect because Medusa 12 and I were now separated laterally and vertically. If the bandit stayed fighting me, Medusa 12 would have an easy time shooting him. Likewise, if the bandit switched to go after Medusa 12, I would be in a position to quickly kill him instead.

The problem was, I was slow coming out of the last fight, and executing the high-G pull straight up into the bandit ran me out of airspeed before the afterburner could reach full power. The bandit flew past me in a dive, and just as my nose touched straight up, the low-speed warning horn sounded in the cockpit. This horn warned me that the jet was reaching a dangerous combination of low speed and high angle of attack, which is exactly the conditions needed to put the jet out of control. When the horn went off, I was turned around in my ejection seat watching the bandit as he flew by, going almost straight down. Now, getting the low-speed warning horn happens all the time when you are dogfighting, so I did what I always had done; I let go of the stick so that the airplane could flop over nose-down and fly again. What I didn't realize right away was that I was literally going perfectly straight up, and the nose didn't have a natural direction to fall in. As soon as I let go of the stick, I keyed the mic and over the radio called, "Medusa 11 is ballistic." This call would let everyone know I was letting my jet fall off unpredictably to one side or another, and to stay out of my way.

But after a few seconds I realized the nose of the jet was staying straight up and not wavering a bit. I saw that the airspeed was indicating zero, and a check of the angle of attack gauge showed it was pegged off the top of the scale. "Hmmmm," I thought to myself, "this looks just like what I saw Friday on my out-of-control flights at Edwards." Using the lessons learned from my flight at Edwards AFB as a starting point, I twisted back around in my seat and studied the flight control surfaces. The horizontal stabilizer was banging trailing edge up and down frantically as the computer tried to make the jet fly again. My immediate concern was to check my altitude. We were fighting over the ocean, and when you are over water it can be very easy to lose track of your altitude and be a lot closer to hitting it than you realize. I looked at the altimeter and saw it was unwinding down through 16,000 feet. Later, when the flight data from the jet was downloaded, I learned that my jet reached zero airspeed going straight up at 18,000 feet, and then slid backwards for about 2,000 feet before anything really exciting happened. In the jet, at that moment, all I knew was that I was dropping tail-first.

Now, jet engines like to have the air come in the front and go out the back. Since I was falling tail first, I was concerned that the engine, running in full afterburner now, would compressor stall and stagnate or even fail with no air coming in the front end. To protect the engine, I pulled the throttle back out of afterburner. It also occurred to me that I was in a jet with a centerline tank, and based on my experiences at Edwards, I knew this could turn into a very violent ride. I reached down and threw the lever that would lock my shoulder harnesses and hold me firmly in the seat. That turned out to be a prudent move, because as soon as the engine spooled out of afterburner the jet flipped violently over onto its back. The jet paused for a second, frozen upside down, and then began to yaw fast to the left in a flat attitude...kind of like a record spinning on a turntable. I glanced at the altimeter again and saw that I was passing 14,000 feet. That was lower than I was comfortable with. I decided that I had given the computer enough time to try and recover, so I engaged the MPO switch and started rocking the nose up and down. When combined with being upside down and spinning like a record, the effect was dizzying! After pitching the nose up and down a few times, the jet abruptly stopped yawing left and began yawing right even faster. In the background I could hear the bandit calling a "Knock-it-off" to stop the fight; he could see that I was out of control. I was doing my best to ignore the fast yaw rate and divided my time between watching the altimeter and timing my stick movements to increase the amount of pitch as the jet began to see-saw nose up and down in bigger oscillations. All of a sudden, the jet did about three un-commanded and very violent rolls and stabilized upright, yawing back to the left again. Looking at the altimeter again, I saw that I was now passing 10,000 feet. The F-16 has a great ejection seat, but the aircraft manual warns you that in an out-of-control situation, you need to eject no later than 6,000 feet to make sure you get out in time. Given that the jet was behaving badly compared to my recent flight at Edwards, I began to think I was going to have to punch out. Just as I was passing 8,000 feet the yaw rate stopped, and on my next pitch oscillation the jet's nose popped straight down. The airspeed began to register, and the angle of attack gauge indicated I had air over the wings again. This was great news, as I was just now approaching 6,000 feet. But I knew I wasn't out of the woods yet. When the F-16 nose first drops after recovering, it is very easy to over-control and depart again. The sensation is kind of like balancing on a ball, and you can feel that every little wobble could cause the ball to roll out from under you. So, with a light touch on the stick, I kept the nose centered straight down until I was passing 250 knots and felt comfortable pulling back out of the dive. Soon I had the jet back flying level at a little less than 4,000 feet above the water.

Given what had just happened, we terminated the mission, and I declared an in-flight emergency. The jet seemed to be flying fine, and my engine gauges didn't show any problems. My wingman rejoined on me and gave my airplane a good look-over. He didn't see any parts missing or other damage. Just to be safe, however, I landed at the nearest suitable airfield, which happened to be Myrtle Beach airport. A few hours later a maintenance team showed up in a couple of busses from Shaw AFB, and I caught a ride back to base on one of them. It took a couple of days for the maintainers to look the jet over, and they declared it was good to go. I went back over to Myrtle Beach a few days after that and flew the jet home. Back then, putting an F-16 out of control was considered an accident, so there was an investigation. The investigating pilots who interviewed me had a good laugh when they found out I had been to Edwards and put an F-16 out of control 13 times three days before my mishap! Reflecting back the whole thing, it really was a stroke of good luck. I am sure I would have recovered the jet without the training at Edwards, but that recent experience really made the whole situation easier to deal with. As I have always said, "It's always better to be lucky than good!"

A Nickle on the Grass

Flying fighter planes is inherently dangerous work, and sadly I have lost many friends over the years. As a rule of thumb, I don't like to tell those stories, but there is one I want to share because the pilot who was killed was one of the best fighter pilots I ever flew with…and a close friend.

I first met Aaron "C-Dot" George at Columbus AFB, in Mississippi, when we were going through fighter lead-in training together. After you graduate from pilot training, if you are going to fly fighter planes, you attend a six-week course flying AT-38Bs to learn the basics of being a fighter pilot. On the first morning of class, C-Dot walked up to me. "Hi! I'm Aaron George," he said shaking my hand, "and I want to be an astronaut on the first mission to Mars!" I laughed, but he was serious. That was the start of a long friendship. C-Dot and I both went to Luke AFB after fighter lead-in and learned to fly F-16s in the same squadron there. We were squadron mates at our next two assignments as well. Our families spent a lot of time together too, and my daughter Katie and his little girl Jessie were the same age and played together every chance they could get. C-Dot hailed from West Virginia, had an advanced degree in aeronautical engineering, and genuinely wanted to be an astronaut and go to Mars. He was an excellent fighter pilot, and it was no surprise to anyone when he was selected to attend the U.S. Air Force Test Pilot School at Edwards AFB, in California. C-Dot knew that most Air Force astronauts had been test pilots, and if being a test pilot would get him to Mars that was what he would become. When he graduated from Test Pilot School me, and another pilot, took a road-trip and attended the ceremony.

One day in the summer of 2001, C-Dot was assigned to fly a test mission in a two-seat F-16. He was testing a Miniature Air-Launched Decoy, or MALD for short. The Air Force was very interested in MALDs at the time, and they were testing lots of different ones designed to look like different airplanes on radar or from an electronic emissions viewpoint. The idea behind a MALD is that you can launch a swarm of them, have them behave like real airplanes flying to a target. Hopefully the enemy's air defenses would be fooled and expend all of their missiles shooting down the MALDs and not the actual friendly fighters. On this particular mission C-Dot had a cameraman in his backseat, and another F-16 flying as both the airplane launching the MALD and as a safety chase-plane. The plan was for the MALD to be launched, and then C-Dot would fly alongside it, while the guy in the backseat filmed it for later analysis. When the MALD launched, the chase-plane would make a 360 degree turn to get out of the way and then provide safety-chase from a few miles behind.

The test flight took place very early in the morning, over one of the valleys that run roughly north to south near Edwards AFB in the Mojave Desert. It was early enough in the morning, and the rising sun had not yet cleared the mountain peaks overlooking the valley, so the valley floor was still in deep shadows. Everything proceeded as planned, and after launching the MALD, the safety chase plane executed his 360 degree turn to get behind C-Dot. As the safety chase plane was completing the turn, he saw a fireball light up the dark valley floor below. C-Dot and his cameraman had crashed, and both were killed. It would be a few weeks before the accident investigation was able to determine what had happened.

The accident report revealed that just after C-Dot launched the MALD, a malfunction occurred, and the MALD deployed a small recovery parachute. This made the MALD drop straight down instead of flying along at altitude. In an effort to keep the MALD in sight, C-Dot whipped his jet into a high-G, left-hand, descending turn. The investigation determined that the combination of the high-G turn and C-Dot looking over his left shoulder, created a physiological phenomenon called G-Excess Illusion. Basically, his body posture in relation to the G-forces and bank angle tricked his brain into thinking he was rolling back to the right, so he subconsciously rolled harder to the left. At the same time, with the valley floor being obscured by shadows from the angle of the rising sun, C-Dot didn't recognize how close he was to the ground in time to recover from the steep high-G spiraling dive. The flight data recovered from the crash showed that when C-Dot did become aware of the dire situation, he correctly rolled the airplane to the nearest horizon and tried to save the jet with a 9-G wings-level pull. Amazingly, the F-16 actually hit the ground and skipped-off the valley floor intact. C-Dot could have probably flown away from ground, and safely ejected. He might have even been able to limp back to Edwards and land. There was, however, a large, house-sized boulder on the valley floor, and moments after skipping off the ground, C-Dot's jet flew right into that big rock.

I was devastated when I heard the news. I had just been out to Edwards AFB a couple of months earlier to attend C-Dot's Test Pilot School graduation, and now he was gone. Another pilot and I drove over from Luke AFB for the memorial service. Guys from all over the Air Force turned up as well, to pay their last respects. When I saw C-Dot's wife we hugged, and I cried like a baby. At the actual memorial service, I was sitting a few rows behind C-Dots kids, and their grief absolutely broke my heart to see. I kept looking at his daughter, who was a good friend of my daughter, and all I could think was, "That could be my Katie one day."

After the memorial service all of the fighter pilots present went over to C-Dot's squadron there at Edwards. We all congregated in the squadron bar and poured tumblers of whiskey. In turn, we each stepped forward and told our best C-Dot story, celebrating a great pilot and friend. One by one we drained our glasses and smashed them against the wall. After the last glass was broken, everyone gaggled outside to throw a nickel on the grass. This is an old fighter pilot tradition from World War 2. It comes from a line in a fighter pilot bar song, "Throw a nickel in the grass, to save a fighter pilot's ass. Throw a nickel in the grass and you'll be safe." From there everyone made their way over to the Officer's Club and drinking continued late into the night. Somewhere north of midnight, I staggered out of the bar and tried to find my way back to the Bachelor's Officer's Quarters where I was staying. Lost, and somewhat foggy from the evening's libations, I eventually found myself out on the base golf course. As I wandered around, I remember thinking that C-Dot was a good man, and that if there is a heaven that's where he had to be. Eventually I laid down on one of the greens and looked up at the stars in the crystal-clear, night desert sky. I drifted off to sleep thinking about how C-Dot wanted to be an astronaut and visit those stars…and maybe that was what he was doing right now.

UFOs

Earlier in this chapter I noted that, besides asking me if I had ever ejected from an F-16, the most frequent question I am asked if I have ever seen a UFO? And the answer is yes. Have I ever seen a UFO that I could explain after the fact? The answer is also yes. But I also have to confess that I have seen what I am convinced were legitimate UFOs doing things I cannot explain. Let me elaborate.

I want to start off by saying that I think the majority of UFO sightings are most certainly airplanes. When you are flying, you are surrounded by lighting conditions, visibility issues, and a variety of other visual illusions that can make an airplane look like something that isn't an airplane. Dozens of times I have been flying and looked over at my wingman, only to be amazed that he looks like a glowing saucer or cigar, or some other distinctly non-airplane shape. I know it is my wingman, flying a mile or so off my wing, but the combination of sun angle, how it reflects off his jet, the clouds or terrain in the background all conspire to make him look exactly like a classic UFO. And every time after flying along for a few more minutes the angles would change, and I could clearly see it was an F-16 again. The phenomenon really is amazing, and if I hadn't seen it dozens of times, and watched as the effect faded, I am sure I would be convinced that I had just seen a UFO.

Over the years I have heard some pilots relay their accounts of seeing things that they could not explain. The most convincing ones I have heard usually involved someone flying at a high altitude, say 40,000 feet or higher, looking up and seeing something flying a lot higher than they were. Perhaps these were indeed other airplanes, or even satellites reflecting the sun. For me, however, these accounts are compelling because the observers are experienced fighter pilots and trained to be excellent observers of what they see in the sky around them. There was one story from a guy I flew with whose story was even more interesting than the rest. It actually happened during the Operation Allied Force. The weather was really bad and the pilot in question was flying an instrument approach back into Aviano Air Base. As he was flying the approach something caught his eye outside the canopy. Looking to his left, he saw several glowing orbs flying along beside his airplane. He said that they were clear, distinct, and right beside his jet. He was adamant that it wasn't some reflection on the canopy or the clouds. This guy was a very experienced pilot that I had known and flown with for many years, and he was definitely not the kind of guy to make something up or exaggerate. He said the glowing orbs flew in formation with him for a while, and then flew up and away to his left out of sight into the surrounding clouds.

My most intriguing experiences occurred in the spring and summer of 1996, when I was stationed at Shaw AFB in South Carolina. We flew a lot of our sorties out over the

Atlantic Ocean, especially our air-to-air training missions. In 1996, our F-16CJs had a radar that was significantly more powerful than the radars in any other variant of the F-16. They gave us the ability to see farther and more accurately than just about any other fighter in the U.S. inventory at that time, aside from the F-15 and the F-14. A lot of our air-to-air training involved scenarios that started out far away from our opponents, and we would work to identify and shoot them beyond visual range (BVR). On several of these offshore training missions I would be working my radar scope, when I would see contacts 60 or 70 miles away. Radars often pick up anomalous contacts, but because they are anomalous, you can't lock them up and track them; they simply disappear when you try. But a few times I was able to lock up these contacts, and when I did, they would behave in the most amazing ways. When I would lock them, I would get a good radar track, showing an altitude in the 20,000- or 30,000-feet range, and traveling at 200 or 300 knots on a steady heading. A few seconds after locking them up, the contacts would turn and climb away at insane speeds, sometimes indicating in excess of 1,000 knots within two or three seconds! The radar would maintain lock for the few seconds it took for the contact to depart outside of my maximum radar range.

I actually filmed these events a few times and showed them to the highly experienced pilots in the squadron. I was new to the F-16CJ and its radar, so when they dismissed the contacts as quirky radar behavior, I believed them. Not only were they more experienced than me by far, but I couldn't think of any airplanes owned by any nation that could accelerate that fast. I continued to see them from time to time in the spring and summer of 1996, but I didn't pay them much attention once they flew off my scope. I saw them again in the 2006-2007 timeframe at Shaw AFB as well.

Fast-forward to 2020, and I saw something that caused me to, in hindsight, re-evaluate what I saw in 1996 and 2006-2007. In 2019 and 2020, the news was full of stories and videos about U.S. Navy pilots who saw and filmed UFOs off the west coast of California. Astonishingly, the Navy confirmed the videos were real! Soon more footage emerged from Navy sightings off the east coast of the United States as well. The news articles were pretty focused on the videos, which were amazing to see. I searched the internet and found an actual video-taped interview with the F/A-18 squadron commander at the center of the controversy. Unlike the print articles, he also described in detail what they saw on their radars, and it sounded exactly like what I had seen. Additionally, the interviews talked in more detail about other Navy pilots who had seen and recorded things off the Atlantic coast of the U.S. One of the commenters noted that the sightings occurred after the F/A-18s had gotten a new software tape for their radar. That rang a bell for me…in early 1996, and several times in the 2006- and 2007-time frame we had received a new software update for the F-16CJ. These kinds of software upgrades frequently included radar capabilities. Soon after the news and videos of the Navy encounters became public, the Department of Defense launched a new investigation into these unknown aerial phenomena. Unsatisfying reports to Congress were also published.

Thinking about all of this, I now believe that that what I saw on my radar back in the day was not a radar anomaly after all, but instead something real. Maybe they were airplanes of some sort operating far from the coast, and when I would lock them up, they would jam or spoof my radar into thinking they were travelling away at high speeds. If that was really what I was seeing, I hope it was a U.S. airplane and not one of our adversaries with such powerful jamming equipment and software. To be honest, I don't think this was the case because I believe I would have known about such a capability, be it U.S. or foreign. Instead, I am drawn to the fact that these Navy pilots not only saw the same things on radar (after receiving new radar software), but that they were able to correlate it to recorded targeting pod footage and visual accounts confirmed by multiple fighter pilots at the same time. Perhaps there was something out there that had remained hidden until it was surprised by new radar capabilities? Does that mean it is an alien spacecraft? I don't know, maybe? But I do know that I saw something I can no longer explain or ignore. So, do I think I have seen UFOs? Yes,   absolutely.

Chapter 16: Epilogue

After returning from Operation Allied Force life didn't slow down at all. I received orders to Luke AFB, Arizona, where I taught new pilots how to fly the F-16. It was a pretty good assignment, but I missed being in a front-line fighter squadron, especially after the September 11th, 2001, attacks. I would get my chance to deploy to Iraq and Afghanistan many times in the years ahead, but it was hard to be on the sideline when it all started. While at Luke AFB, I was selected to be a Core Development Pilot on the F-35 program. It was really amazing to help develop America's newest and most capable fighter. I spent a lot of time at the Lockheed-Martin factory in Fort Worth, Texas, working with a great group of engineers and the other Core pilots developing the Pilot-Vehicle-Interface and Strategic Attack concept of operations.

After my tour at Luke was done, I attended Air Command and Staff College (ACSC) at Maxwell AFB,

Alabama. ACSC was a 10-month program, and in addition to picking up a second master's degree while there, I had the opportunity to learn just how much I love flying; while at ACSC pilots don't get to fly, and it just about killed me to sit in a classroom knowing my fighter pilot brethren were tearing things up overseas. My follow-on assignment from ACSC was to a staff position at 9th Air Force/U.S. Central Air Forces (USCENTAF). Despite not wanting to go to a staff job, I ended up having a pretty good time. I spent almost the entire 18 months of that assignment in the Middle East, got to play Marine and Army Soldier in Afghanistan and Iraq respectively, and actually learned a lot about how air wars are run by the Generals in charge. I also was promoted to Lieutenant Colonel, completed Air War College by correspondence, and came out on the list of folks eligible to command a squadron.

From my staff tour I went back to Luke AFB briefly to requalify in the F-16. Fortunately, I was out of the cockpit less than 3 years, so I didn't have to go back to the centrifuge! I returned to Shaw AFB, South Carolina, and served as the Chief of Wing Safety, before serving as the Director of Operations for the 55th Fighter Squadron. I was next selected to be the Commander of the 20th Operations Support Squadron, and then to be the Deputy Operations

Group Commander at Shaw AFB before retiring in 2008. The years since I retired have also been busy. I earned a PhD, continued to fly F-16s as a civilian, and worked as a test pilot for one of the major aerospace manufacturers. There wasn't enough adrenaline in test pilot work, so I returned to combat as a private military contractor, flying little propellor airplanes on high-risk security missions in some of the world's most dangerous places. Obviously, there is a lot I could unpack between the end of Operation Allied Force and today, but for now I am happy to have told the story of what it was like to be a combat fighter pilot over Kosovo!